Who is Conscious?

Who is Conscious?

A Guide to the Minds of Animals

MARIAN STAMP DAWKINS

OXFORD
UNIVERSITY PRESS

Oxford University Press is a department of the University of Oxford. It furthers the University's objective of excellence in research, scholarship, and education by publishing worldwide. Oxford is a registered trade mark of Oxford University Press in the UK and in certain other countries.

Published in the United States of America by Oxford University Press
198 Madison Avenue, New York, NY 10016, United States of America.

Library of Congress Cataloging-in-Publication Data
Names: Dawkins, Marian Stamp author
Title: Who is conscious? / Marian Stamp Dawkins.
Description: New York, NY: Oxford University Press, [2025] |
Includes bibliographical references.
Identifiers: LCCN 2025035834 (print) | LCCN 2025035835 (ebook) |
ISBN 9780197818626 hardback | ISBN 9780197818633 epub | ISBN 9780197818657
Subjects: LCSH: Consciousness
Classification: LCC BF311 .D389 2025 (print) | LCC BF311 (ebook)
LC record available at https://lccn.loc.gov/2025035834
LC ebook record available at https://lccn.loc.gov/2025035835

DOI: 10.1093/9780197818657.001.0001

Printed by Integrated Books International, United States of America

The manufacturer's authorized representative in the EU for product safety is Oxford University Press España S.A. of Parque Empresarial San Fernando de Henares, Avenida de Castilla, 2 – 28830 Madrid (www.oup.es/en or product.safety@oup.com). OUP España S.A. also acts as importer into Spain of products made by the manufacturer.

Preface

This book grew out of the frustration of seeing so many claims about animal consciousness going unchallenged. Tentative suggestions about which animals might have conscious experiences were being treated as more definite than the evidence seemed to justify. No one seemed to be putting an alternative view.

So here are some alternative views, intended to be a positive contribution to the ongoing debates about which animals are conscious. I offer them in the spirit that we are most likely to reach a valid conclusion if the evidence we use has survived questioning and scrutiny. Doubting and being critical are essential for eliminating unreliable claims. We owe it to animals to make sure that the case for their welfare is based on the best possible evidence about their sentience.

The book is for everyone interested in animals and what might be going on in their minds.

Marian Stamp Dawkins
Oxford December 2024

Acknowledgements

I am very grateful to many people who, even though they may not have realised it, have made this book possible by their support, encouragement, ideas, and criticisms. They include Manuel Berdoy, Jonathan Birch, Alan Grafen, Joe LeDoux, the late (and greatly missed) Aubrey Manning, Georgia Mason, and the numerous students who have shared my interest in animal minds. I am particularly grateful to Edmund Rolls for sharing his inspiring views on consciousness, for commenting on sections of the book and still encouraging me to keep writing.

Contents

1
Introduction

For many people, the question of animal consciousness has already been answered. Yes, of course, animals such as dogs, whales, chimpanzees, and birds have conscious experiences. Of course, they can feel pain and pleasure, happiness, and fear. To argue otherwise would not only be heartless, but it would also fly in the face of the increasing number of claims that consciousness is not just a property of human brains but is, rather, an evolutionarily ancient trait that evolved millions of years ago and is now widespread in the animal kingdom. There was certainly no doubt in the minds of a group of eminent scientists who, in 2012, issued the Cambridge Declaration on Consciousness.[1] They stated it as a matter of scientific fact that many non-human animals including all mammals, birds, and octopuses have what they called the 'neurological substrates' for generating consciousness. And, reflecting the public mood, animal consciousness is also now recognized as a fact in law across the world, with the European Union and the UK Parliament both declaring that animals are sentient as the basis for their legislation.[2]

So, if animal consciousness is now accepted as a statement of the obvious by public opinion, supported by scientific evidence and regarded as a proven fact as far as the law is concerned, why is there any need to continue to ask questions about it? Wouldn't this just be giving legitimacy to people who want to deny that any animals other than humans are conscious and so risk animals being treated worse than ever? Wouldn't it be better for the animals' sake just to go with the flow and accept what evidence there is, imperfect

[1] Cambridge Declaration (2012)
[2] Martinez and von Nolting (2023)

though it may be, that many animals have conscious experiences much like our own? Having spent much of my career on animal welfare at a time when it was not nearly as fashionable as it is now, shouldn't I just be glad that the last fifty years have seen such a sea change in attitudes to animals—undoubtedly for the better—and not continue to raise difficult questions that might endanger that progress?

There are two good reasons why it is now more important than ever to keep asking questions about animal consciousness. The first reason is practical. There is a widespread view that mammals, birds, and even all other vertebrates are capable of conscious experiences, but there is a lot more doubt about insects, crabs, snails, and other invertebrates. In fact, there is no consensus at all about where we should draw the line between animals that have conscious experiences and those that do not. If you look at the various theories that have been put forward, you can pretty much take your pick. At one extreme are theories that see only humans and apes as having conscious experiences.[3] Slightly less exclusive are theories that regard all mammals and birds[4] or mammals and birds and even reptiles as conscious beings while excluding fish, amphibians, or any invertebrates.[5] Many scientists now say that there is good evidence that fish also have the capacity for conscious experiences[6] while others disagree.[7] There are many people who include octopuses and other cephalopods in the category of conscious beings[8] and increasing numbers are now arguing that flies and crabs should be included too.[9]

There are those who think that even plants may be conscious[10] and yet others who think consciousness is a property possessed by

[3] Macphail (1987); Bermond (2001)
[4] Panksepp (1998); Humphrey (2022); Seth et al. (2005) Boly et al. (2013)
[5] Cabanac et al. (2009)
[6] Mashour et al. (2013); Braithwaite (2010); Sneddon (2015)
[7] Rose et al. (2014); Key (2015); Hart (2023); Diggles et al. (2024)
[8] Godfrey-Smith (2016); Crook (2021)
[9] Barron and Klein (2016); Bronfman et al. (2016); Crump et al. (2023)
[10] Chamovitz (2021); Segundo-Ortin et al. (2023)

all living things[11] and even the whole non-living world, an extreme view called Panpsychism.[12]

This is not a good state to be in. We need to know where to draw the line between animals that have conscious experiences and those that do not because this has such major implications for the way we treat them. Conscious beings have a special status. The capacity to feel pleasure and pain lifts something from being an object to an entity that matters, ethically, in a unique way. Trees, habitats, and works of art are valuable and worth preserving for a variety of perfectly valid reasons. But animals that are capable of conscious experiences have an additional reason for our ethical concern. They matter more because they consciously experience pleasure, pain, and suffering.[13] So we need to keep asking questions about consciousness because we need to know whether we should be giving as much ethical consideration to the fleas on a dog as to the dog itself, and to the mites on a chicken as to the chicken itself. Giving legal protection to mosquitoes, flies, or beetles would have massive implications for agriculture and human health and would cause a complete rethinking of our attitudes to the animal kingdom in such areas as pest control, scientific experiments, and food. So even for those who think the question of animal consciousness has already been answered, there is still at least one further unknown: Who is conscious?

There is, however, a second reason for wanting to keep asking questions about animal consciousness that is perhaps even more important. This is that the issue of which animals are conscious is only one of many unanswered questions that can be asked about the greatest unsolved mystery in biology. Despite all that has been researched and written about consciousness, we still do not understand how it arises or, to be more specific, how the lump of nervous tissue that makes up our own brains could possibly give rise to the

[11] Margulis (2001)
[12] Chalmers (2016)
[13] Singer (1975); Midgley (1983); Rollin (1989); Mellor (2019)

awareness we all experience every day. It seems such an extraordinary thing. About 1.4 kg of grey, wet-looking cellular mass gives rise to the conscious experiences that make us all what we are. How does a brain do that? And which other brains can do that too?

Although we now know a great deal about how human brains are constructed and even have imaging technology to show how living brains change as people think, remember, or respond to stimuli, we still do not understand how that brain activity gives rise to conscious experiences. Not understanding that in ourselves makes it doubly difficult to know what is going on in the brains of other animals, particularly those whose brains may be very different from our own, such as those of birds and octopuses. Without knowing how consciousness arises in ourselves—that is, what makes a brain have conscious experiences at all—we don't know what to look for when we turn to the variety of other brains that exist in the animal kingdom. And there are not just animal brains to think about.

Although it is not the intention of this book to cover the issue of consciousness in machines, the existence of ever-cleverer robots and artificial intelligence raises the possibility that consciousness may not even need a brain at all, at least not one made out of nervous tissue.[14] It might also occur in non-living things such as computer circuits or clouds of gas. If it does, this is directly relevant to our understanding of where consciousness comes from and the circumstances that give rise to it. It means we may have to expand our ideas of who is conscious and who is not. The questions surrounding animal consciousness thus lead us inevitably to a whole range of other, even more profound, questions to which we do not yet have answers. What is consciousness? How does it arise? What is it for? It is these and other questions that form the focus of this book.

We are—because it affects so many of our ethical values—almost desperate to have a theory of what makes a conscious mind. We

[14] Butlin (2022)

want to know whether this or that animal is capable of conscious experiences because that will affect how we treat them, whether we choose to eat them, and how much priority we give them in relation to human lives. We want to know the precise time when a human embryo becomes conscious because that will affect our attitude to abortion. We want to know whether machines that exhibit human-like behaviour also have human-like conscious feelings because that will determine how the machines of the future should be integrated into human society.[15] We feel we must conquer consciousness now because we are under pressure to make important ethical decisions now.

Unfortunately, this urgent desire to solve the mysteries of consciousness is itself a major blockage to understanding it. It stops us from asking questions about what we don't know because we fear that in doing so, we might undermine the ethical views we have already decided on. Once people have decided that they have a sufficient knowledge of consciousness to be able to build it into, say, laws and regulations about animal welfare, they no longer wish to be reminded about how imperfect that knowledge is because to do so might challenge their world view or cast doubt on a cause they have given their lives to. They fear that it might suggest that animals were not sentient or even lead to animals being mistreated. Just asking questions or pointing out what we don't know can then seem like a threat to deeply held ethical values.

In reality, however, there is no such threat. All we have to do is to accept that the standard of evidence about consciousness we need to make an ethical decision is lower than the standard of evidence we accept if we want to say that we fully understand consciousness as a scientific phenomenon. I can accept that a dog is conscious, behave towards it as a conscious being, believe it to be ethically wrong to cause it harm and support laws to protect its welfare while at the same time still saying that, from a scientific point of view, I

[15] Shevlin (2021)

do not know for certain whether it is conscious because I don't understand enough about consciousness to be able to do so. It is important to add that I would say exactly the same about other people. I would argue that anyone who is reading this book is certainly consciously aware, but I would also argue that scientifically there is a lot I do not understand about what gives rise to their conscious state. I am even more baffled by my own consciousness. I experience that with a certainty that allows of no doubt, but I still don't understand it and I still want to find out more about what makes me into the conscious being I am sure I am. In other words, we can be ethically certain at the same time as being scientifically uncertain. We can have deeply held beliefs about how we ought to behave and still want to raise questions and doubts about the basis of those beliefs.

The problem is that this distinction—between the necessarily incomplete evidence about consciousness that most people are prepared to accept when they make decisions about how to behave or pass laws and the much more rigorous thinking we need to apply when tackling the mystery of consciousness itself—can so easily become blurred.[16] And in blurring that distinction, our thinking about consciousness itself can become less rigorous. The sensible, pragmatic step of basing ethical decisions on evidence that is imperfect but 'good enough' has the unintended consequence of making it more difficult to subsequently investigate consciousness objectively. We become trapped in our own psychological desire to be right. Our views on highly emotive ethical issues (such as animal welfare) depend on what view of consciousness we have already adopted and therefore make it less rather than more likely that we will be able to confront what we don't know.

This book is about doing just that—confronting what we don't know about consciousness. It is about asking questions about animal minds, but it is certainly not about denying consciousness in

[16] Dawkins (2022)

animals or putting forward a particular view of which ones have it. For if we really want to know who is conscious, we cannot avoid asking how good the evidence really is. If all we wanted to do were to make reasonably plausible ethical decisions and sensible workable laws, we could simply accept the evidence at face value. As Jonathan Birch[17] has argued, we don't need to have solved the hard problem of consciousness before deciding that we have enough evidence to bring animals under the protection of the law. We can blur a few boundaries and sidestep any number of issues and still be reasonably confident that we have made the right decisions, or at least decisions that are workable and not too outrageously wrong.

However, to really understand consciousness and to come to valid scientific conclusions about which animals might have it, we must try not to blur any issues or sidestep anything. A scientific approach to consciousness means raising doubts and asking questions. It means challenging certainties and being prepared to go to places that many people do not want to visit—that is, deliberately facing up to what we don't know.

Now, unknown places are best navigated with some sort of guide or at least a map showing where the main hazards are. The aim of this book is therefore to share with you some tips that I have found useful in thinking about consciousness in animals, in the hope that you will find them helpful too. The book is not a complete account of animal minds because there are so many things we still don't know, and a comprehensive account cannot yet be written by anybody. And it certainly will not tell you how to treat animals or whether we should eat them or hunt them or keep them as pets. There will be no attempt to convince you of any particular view of animals or how they should be treated. I hope, therefore, that you will find it useful whatever your current views—whether you have already made up your mind about animal consciousness or whether you are genuinely unsure what view to take. The

[17] Birch (2017); Birch (2024)

issues are confusing. What we are trying to explain is mysterious. The evidence often looks more convincing than it turns out to be on closer inspection. But that's what makes this whole field so endlessly fascinating—and so worth pursuing.

There is a danger, however, not so much in asking questions but in being misunderstood as to why we are asking them. So, before we get down to asking who is conscious, I need to be quite clear what this book is not about. It is not about denying that animals are conscious. Subjecting a body of evidence to scrutiny does not carry the implication that such evidence is wrong. Quite the contrary. Good evidence survives scrutiny and is all the better for it, so questioning the evidence for animal consciousness—if it is that good—should actually improve the case for consciousness in beings other than humans. As the next chapter explains, the case for taking animal welfare seriously will be stronger if it is built on solid evidence than if it relies on assumptions that can easily be shown to be unreliable. The purpose of probing the evidence is thus not—repeat, not—to demolish the case for animal consciousness. It is to take a few steps towards a fuller understanding of something quite extraordinary that we currently do not understand—consciousness itself.

2

Questioning consciousness does not threaten animal welfare

During my working lifetime, attitudes to studying consciousness—human as well as animal—have undergone a remarkable change, from complete avoidance to an almost unthinking acceptance. In 1951, Niko Tinbergen, Nobel Laureate and often described as one of the founding fathers of the study of animal behaviour, wrote in his landmark book *The Study of Instinct*:[1]

> *Because subjective phenomena cannot be observed objectively in animals, it is idle either to claim or deny their existence*

He then elaborated further (p.5):

> *Hunger, like anger, fear, and so forth, is a phenomenon that can be known only by introspection. When applied to another subject, especially one belonging to another species, it is merely a guess about the possible nature of the animal's subjective state.*

Tinbergen, who also happened to be my doctoral supervisor, was quite open to the idea that animals *might* be conscious. He just thought it was not possible to tell if they were. It was therefore not a proper area of study for anyone wanting to follow a serious scientific career path. I was warned off.

Then came the dissenting voices. In 1964, Ruth Harrison published her book *Animal Machines*,[2] in which she documented the

[1] Tinbergen (1951) pp.4–5.
[2] Harrison (1964)

way many farm animals were kept. She shocked the world with her revelations of what 'factory farming' was doing to animals. With an ability to generate publicity that still appears impressive in our age of social media, her views were splashed all over the newspapers and, most importantly, led to the setting up of a UK Government Committee, the Brambell Committee, to look into the animal welfare implications of modern farming methods. For the first time, battery cages for laying hens, crates for sows giving birth, and other new developments in farming came under public scrutiny.

One member of the Brambell Committee, the distinguished Cambridge biologist W.H. Thorpe, was asked to give a scientific assessment of whether these practices caused suffering to farm animals, thus forcing him to take a view on the issue of animal consciousness. Do animals consciously experience pain and distress, or do they not? Thorpe broke with the prevailing scientific view of the time that such a question is unanswerable and gave a very positive answer. In an essay published as an appendix to the main Committee Report,[3] he suggested that there should be a major research project dedicated to just this question. He ended his essay by quoting from a recent lecture given by Lord Brain:

> *I personally can see no reason for conceding mind to my fellow men and denying it to animals … … … …. I at least cannot doubt that the interests and activities of animals are correlated with awareness and feelings in the same way as my own, and which may be, for ought I know, just as vivid.*

Thorpe argued that animal suffering should now be a priority for research. The study of animal welfare—including the issue of what animals feel—should no longer be considered beyond respectable science.

[3] Thorpe (1965)

As we will see in more detail later in this book, the 1960s and 1970s saw a great expansion in the number of studies on consciousness, in both humans and non-humans. For animals without the language to communicate what they were feeling, there was a realization that behaviour might be the key to understanding their inner subjective worlds. The way that an animal behaved, and in particular how it communicated to others, was seen as key to accessing what it was consciously experiencing. After all, actions speak louder than words, even for us, and it was reasonable to argue that animals, like us, could express their emotions through gestures and sounds that do not need words. Animal behaviour, correctly interpreted, appeared to open a window into the minds of animals.[4] Animals that made choices and could show a preference for one environment over another were expressing their opinion just as surely as if they had articulated it in fully formed sentences.[5]

Of course, the argument that animals do consciously experience emotions such as fear, anger, disgust, boredom, or pleasure was not watertight since no one could be sure that there were conscious minds behind the observable behaviour but the balance of evidence was shifting. The more we learnt about behaviour and the remarkable things that animals can do, such as build dams, coordinate their hunting, warn others about predators, and anticipate what another animal was about to do, the more plausible it seemed. The idea that they were just 'going through the motions' and not consciously experiencing anything at all remained a possibility but an increasingly unlikely one. Besides, repeatedly saying that we did not know for certain that animals had conscious experiences did not seem particularly helpful and, indeed, could undermine the real progress that was being made in improving animal welfare.

[4] Griffin (1976); Griffin (1984); Burghardt (1985)
[5] Hughes and Black (1973); Dawkins (1990)

That, at least, was the view I held for many years. I even wrote a book about it.[6] I was aware that the evidence for animal consciousness was not always as strong as it was sometimes made out to be, but I thought the best way forward was to make the strongest possible case for consciousness in animals while at the same time making it clear that the evidence was still very tentative.[7] What made me change my mind was the way in which people—including scientists—began talking as if we had far more definite knowledge than we really have.

> '*Animals display flexibility in their behaviour patterns, and this shows* that they are *conscious and passionate and not merely "programmed" by genetic instinct to do "this" in one situation and "that" in another situation*', wrote Marc Bekoff '*My colleagues and I no longer have to put tentative quotes around such words as happy or sad when we write about an animal's inner life. If our dog, Fido, is observed to be angry or frightened, we can say so with the same certainty with which we discuss human emotions.*'[8]

Virginia Morrell was equally certain about consciousness in animals. She began her book *Animal Wise* (2013) with the words:

> '*Animals have minds. They have brains and use them, as we do: for experiencing the world, for thinking and feeling, and for solving the problems of life every creature faces.*'

The Cambridge Declaration on Consciousness (2012), which was mentioned at the beginning of the last chapter, set out a quite unequivocal view of animal consciousness with the backing of an international group of prominent cognitive neuroscientists,

[6] Dawkins (1993)
[7] Rowan et al. (2021)
[8] Bekoff (2007) p.31

neurophysiologists, neuroanatomists, and computational neuroscientists. This group had gathered at Churchill College at the University of Cambridge and began by stating their view that humans are not unique in having conscious experiences. They then went on to declare that many animals, including octopuses, have the same neurological substrates for generating consciousness as humans do. While this is a plausible view, to declare it as fact goes far beyond what the scientific evidence currently shows us. We do not know what the 'neurological substrates for generating consciousness' are in any animal, human or non-human[9] and currently have no way of showing that these are present in octopuses.

Gone was the caution I thought should accompany any statement about animal consciousness. In its stead was a certainty about the minds of animals that implied the existence of a large body of established scientific facts with no room for any doubt or questioning. Indeed, doubt and questioning were roundly criticized, as if anybody who dared to go down that path must also believe that '*animals are just objects, activated by responses to environmental stimuli*' as Jane Goodall[10] put it, or that they saw animals as zombies or robotic machines '*moving through life like the half dead*', to use Virginia Morell's even more damning words.[11]

Worse than that, doubters and questioners were said to be impeding progress towards improving animal welfare. '*After all*', continued Jane Goodall, '*it is easier to do unpleasant things to unfeeling objects—to subject them to painful experiments, raise them in intensive factory farms, and hunt, trap, eat, and otherwise exploit them—than it is to do these things to sapient, sentient beings*'. Not only was asking questions about animal consciousness scientifically out of date, it encouraged the ill treatment of animals. Marc Bekoff even called it 'Dawkins' dangerous idea.*

9 Koch (2004); Blackmore and Troscianko (2018)
10 Goodall (2007) p.xiii
11 Morrell (2013) p.1
* And he meant me, not Richard. The 2013 web post has subsequently been removed.

Now, I have a lot of sympathy with the idea that there is a connection between views about animal consciousness and views about how they should be treated. Sentient beings—that is, those capable of feeling pleasure, pain, and suffering—do have a special moral status that sets them apart from valuable insentient entities such as trees or works of art.[12] But I felt there was something wrong with the idea that all doubt and questioning about animal consciousness should therefore come to a stop. There was something profoundly disturbing about the idea that just because many people had decided that animals were definitely conscious, that was the end of the debate. Even more disturbing was the idea that it would be easier to convince legislators to enact laws on animal welfare if doubts about animal consciousness were minimized or conveniently forgotten.

The moral argument for stating with certainty that animals are conscious was that this would improve the way animals were treated. But the idea that this should override the right to say that we are still very ignorant about any kind of consciousness—animal or human—did not seem right to me either. Above all, I thought it was important, as a matter of scientific integrity, to point out where conclusions are still provisional. If something is plausible but still uncertain, it should be presented as plausible but still uncertain. Even when many people are convinced of one particular point of view, questions about it should still be allowed and, indeed, encouraged. If they were not, we would miss out on the chance of finally understanding the greatest unsolved mystery in the whole of biology—conscious experiences.

My problem was how to reconcile these two apparently incompatible goals. Improving the welfare of animals had been a long-term aim of mine, and it seemed to demand emphasizing how much evidence we now have for animal consciousness and playing down the doubts and uncertainties. However, for as long as I

[12] Singer (1975); Midgley (1983); Rollin (1989)

can remember, I have also wanted to understand the nature of consciousness and who has it, and this seemed to mean doing the exact opposite. It meant asking questions and emphasizing how much we still don't understand as a preliminary step to understanding things better. It meant raising doubts.

Unfortunately, doubt frequently gets a bad press. It is too often associated with undermining, demolishing, or destroying something or someone. Doubts about the efficacy of a vaccine, about someone's reputation, or about the authenticity of a painting cause no end of problems. But doubt has a very positive role too. The Catholic Church used to appoint a 'Devil's Advocate,' whose job it was to find every possible reason why someone should *not* be made a saint. The Devil's Advocate had to look at every aspect of a person's life and cast every possible doubt on their suitability for sainthood. If the person's reputation survived this critical scrutiny, they really must have led a saintly life. Having doubts cast on their life and then surviving and showing the doubts were unfounded left their reputation stronger than if no such doubts had been raised in the first place.

Similarly, in science, a theory that has been doubted, tested, criticized, tested again, and survives with its predictions upheld has much greater plausibility than a brand new theory that has not been put to the test. If nobody has doubted a theory and nobody has tested it, nobody should believe it. Despite the widespread idea that scientists spend their time showing that their theories are true, the best test of any theory is for many people to try and show that it is not true, or, more accurately, that it makes predictions that turn out not to reflect what happens in the real world. Good theories 'stick their necks out'. They make unlikely predictions that other theories do not. They invite their own destruction. If, despite everyone's best efforts and their most critical doubts, the theory's predictions still seem to match what is actually happening, then the theory can be provisionally accepted. Of course, it is always possible that sometime in the future, someone else will raise another doubt and suggest another prediction that is not borne out in practice, and so the theory will be discarded. This casting out of theories whose predictions fail does not always work out quite so neatly in practice, as scientists are people too, and they can become very attached to their theories and may be reluctant to discard them even when the evidence goes against them.[13] But throwing out theories whose predictions do not match the real world is the way things are supposed to work and, ultimately, the way science makes progress. Doubt—and the new ways of thinking it gives rise to—plays a vital part in the gradual uncovering of the best version of the truth we so far have.

I saw no reason why the same should not be true in the way we set about understanding consciousness, and in particular, how we evaluate evidence about its possible emergence in animal minds. It seemed almost paradoxical that concerns about animal welfare, which rely so heavily on the assumption that many animals are conscious, should stand in the way of critically evaluating the evidence for animal consciousness itself. If the evidence is that good, animal welfare will benefit from further investigation and from someone pointing out the flaws in any arguments that have been used to support it. Like a true saint, it will emerge vindicated and even more convincing.

A possible way of avoiding this dilemma began to emerge from a debate I had in Oxford in the summer of 2018 with Jonathan Birch, a philosopher at the London School of Economics. We initially appeared to be saying very different things, but a member of the audience stood up and said he was surprised at how little difference there really was between us. And it was in one of Birch's papers, titled 'Animal Sentience and the Precautionary Principle' (2017), that I subsequently found a possible solution to my dilemma.

In his paper, Birch acknowledged the problems that exist with knowing which animals are conscious but then goes on to make the case that we should not be too concerned about this lack of

[13] Goldacre (2008); Bishop (2019); Harford (2020)

knowledge, and we certainly should not require absolute certainty that a species is sentient before giving it legal protection. He argued that the dangers of mistakenly overlooking genuine feelings of pain and suffering in conscious organisms are so great that they far outweigh the opposite danger of assuming consciousness in organisms without it. If there is substantial, albeit incomplete, evidence for consciousness in an animal, then, he suggests, consciousness should be assumed. He proposed adopting what he has called the Animal Sentience Precautionary Principle, or ASPP. The ASPP is as follows:

> *Where there are threats of serious negative animal welfare outcomes, lack of full scientific certainty as to the sentience of the animal in question shall not be used as a reason for postponing cost-effective measures to prevent those outcomes.*

In other words, Birch's precautionary principle directs us to proceed with action, such as passing animal welfare legislation, without waiting until we have full scientific knowledge of animal consciousness. If we have reasonable evidence that an animal is sentient, we should just go ahead and pass the laws anyway.[14]

Here was my solution. All I had to do was accept that ethical and scientific decisions require different standards of evidence. There might be a case for lowering the standards of evidence needed for making ethical and legal decisions about animal welfare, but to really understand consciousness itself and to come to valid scientific conclusions about which animals might have it, we have to do the opposite of what Birch was advocating for getting legislation onto the statute books.[15] We have to face the hard problems of consciousness head-on, not side-step or avoid them. Glossing over objections or minimizing criticisms is the very last thing we should be doing.

[14] Birch (2024)
[15] Dawkins (2022)

So, I concluded, improving animal welfare does not have to be at odds with questioning what we know about animal consciousness. We can support the importance of good animal welfare and even believe that it is very likely, but not certain, that many animals are consciously aware. But that does not prevent us from wanting to find out more about what consciousness is and who has it. In fact, we are liberated—free to ask questions, free to test the evidence to see how good it really is, free to doubt.

I can also accept that not everyone will want to do this. Many people prefer certainties. They want the science to be settled. They want explanations that are true and will not change. They particularly want to know with absolute certainty which animals are conscious because, once they are sure they know, this will determine their idea of how they conduct their lives—what they buy, what they eat, what they wear, what they value, even the way they see themselves as a person. The idea that what appears to be 'true' might turn out to be just the best explanation available at the time and liable to shift as more evidence comes in is uncomfortable, unsettling, and, to many people, deeply unsatisfactory.

Deliberately doubting is psychologically difficult to do, and not everyone is prepared to do it. But that is what I am asking you to do if you continue to read this book. I have written the book on the basis that the welfare of animals is an important ethical goal for a civilized society and that achieving that goal may sometimes have to involve taking action on less secure evidence than we would like. But I am also asking you to accept that there is a red line over which we should not cross. That red line is that we should never pretend that the evidence is better than it is. We should never use our current views on animal welfare to claim that the scientific assessment of animal consciousness is now settled and that there is no room for further questioning or doubt.

You may agree with Birch that the ASPP is, under some circumstances, a useful way of getting animal protection into law, but I hope you will agree with me that it is not the way to achieve true scientific understanding. It is the very opposite of what we should be doing to fully understand consciousness in ourselves and in other animals. Science does not need the precautionary principle. The hard problems of consciousness—such as when it happens, what gives rise to it, and who has it—are not going to be solved by making assumptions ahead of time or by being constrained from asking certain questions because they might upset someone's moral attitude towards animals or suggest that a law might have been passed without sufficient background evidence. The really hard problems are going to be solved by raising doubts, by questioning, by being critical of evidence, and then seeing where all that might lead, in the hope, of course, that the temporary discomfort of doubting will eventually lead to the reward of a greater understanding of consciousness itself.

The book is intended to make such doubting a positive experience and a real adventure into the minds of animals. Having reassured you, I hope, that wanting to question what we know about animal consciousness is perfectly compatible with the belief in the importance of animal welfare, we can now turn to some of the best ways of looking at the evidence and evaluating the different ideas that people now have about animal consciousness. We will start with what should be an easily accomplished preliminary first step but turns out to be a major issue in its own right—the little matter of defining what consciousness actually is.

3

Flirting with consciousness

What exactly is consciousness? We all experience it every day of our waking lives, but going from what we know from our own experience to giving a rigorous definition is extraordinarily difficult.[1] Consciousness is clearly different from just being awake. You would react quite differently to a surgeon's knife depending on whether you had been made unconscious by an anaesthetic or had just gone to sleep. People often try to describe consciousness as 'being aware' or having a 'subjective feel' and then give examples such as seeing a red light, feeling an intense pain, or worrying about the future. This pointing to examples is just about the closest anyone comes to defining consciousness, but it only highlights the problem of trying to give a single, unified definition. There seem to be so many different states that we describe as 'conscious'—everything from having long-term financial worries to the momentary pleasure of the taste of food and from the ongoing pain of a headache to a subliminal view of a visual image. Not only do these conscious experiences differ in how long they last, but they also feel different. Pain does not feel the same as disappointment, even though both fall into the category of negative experiences—things we would prefer not to happen. Conversely, the pleasure of drinking when you are thirsty does not feel the same as the experience of being smiled at or a sense of achievement in solving a puzzle, even though they are all positive in the sense that we like and try to achieve them.

Consciousness is clearly not just one type of experience, and as we go through this book, we will need to be open to the possibility that different animals may have very different conscious

[1] Vimal (2009); Blackmore and Troscianko (2018)

experiences, some unlike any that we humans know about. Furthermore, although many people have assumed that it will be the cleverest animals that are most likely to be conscious, there is no reason to assume that they will be the ones most likely to feel basic sensations such as pain. You don't have to be particularly clever to feel hungry or in pain, so there is no necessary link between conscious pain and conscious intellectual ability. As Jeremy Bentham put it in an often-quoted sentence about how we should treat animals: 'The question is not *Can they reason?*, Nor *Can they talk?*, but *Can they suffer?*' (1789). As long ago as 1789, Bentham was making the case for which animals we should care about. He argued that the capacity to suffer was what should make the difference as to how we treat them, not how clever they are or whether they can talk to us in our own language. But in making his case, he was also making a useful distinction between the capacity to feel pain, pleasure, and suffering on the one hand and abilities that are more cognitive, such as thinking, reasoning, and talking, on the other. This is an important point. There is no reason to suppose that these are inevitably linked and that it is only the clever animals (or clever machines) that consciously feel pain or consciously suffer.

The variety of experiences that people describe as 'being conscious' explains the widespread use of another word, 'sentience', that is in common use. Sentience refers to a subset of conscious experiences—the basic ones, such as feeling pain or being aware of a sensation—rather than the more intellectual conscious experiences, such as worrying about the state of the world. Thus, the recent UK and European legislation on animal welfare specifically refers to 'sentience' rather than consciousness, implying that animals, including octopuses and crabs, have at least the very basic elements of consciousness—pain, pleasure, and suffering—but leaving it open as to what other conscious experiences they might also have.

You may also come across the term 'phenomenal consciousness'. The widespread use of this word can be traced back to an influential paper by the philosopher Ned Block in which he distinguished between *phenomenal consciousness* (the whole range of our conscious feelings and experiences) and what he called *access consciousness* (a narrower range of experiences that are associated with reasoning, putting our thoughts into words and rationally guiding our actions).[2] The implication of his distinction is that we can be phenomenally conscious of much more than we can describe because only a subset of our phenomenally conscious experiences make it through to access consciousness where they can be reported on. However, this is a highly contested view, with many people seeing the phenomenal/access distinction as unhelpful, unproven, or even not really a distinction at all.[3] For this reason, the term *access consciousness* will not be used in this book from now on. We will be focussing instead on consciousness in the phenomenally conscious sense—the feeling, experiencing, pain-that-actually-hurts type of conscious awareness—but at the same time we will keep in mind that there may be different underlying explanations for different categories of consciousness, such as feeling hungry or enjoying poetry.

Phenomenal consciousness—awareness—is the really hard problem. David Chalmers put it this way:

> *even when we have explained the performance of all of the cognitive and behavioral functions in the vicinity of experience—perceptual discrimination, categorization, internal access, verbal report—there still remains a further unanswered question: Why is the performance of these functions accompanied by experience?.... Why doesn't all this information processing go on 'in the dark', free of any inner feel?*[4]

[2] Block (1995)
[3] Tye (2017); Carruthers (2019)
[4] Chalmers (1996)

Chalmers is here making the essential distinction between information processing functions such as memory and perception on the one hand and the conscious awareness of these functions on the other. His point is that even if we could fully understand how complex information processing works in the brain, that would still leave us ignorant of why such processing couldn't work without consciousness—why, as he put it, it couldn't just go on 'in the dark'. The real problem of consciousness (part of what he famously called the 'hard problem') is not that we can remember things, recognize faces, or make decisions. It is the element of awareness, of feeling, that may (or may not) accompany these processes. So what we want to know about animals is not just whether they can remember things and recognize objects and make decisions but whether these processes go on 'in the dark' or whether for them, too, there is the light of awareness.

We should not underestimate the difficulty of finding this out. And one of the major obstacles in our path is the language we use. The words we use to talk about consciousness can influence the very core of what we think it is and who might have it. English speakers use the same words—memory, perception, emotion, fear, decision-making, and so on—to mean *both* the information-processing the brain does *and* the conscious feelings that can accompany that processing. We blur the vital distinction that Chalmers was making, often without realizing it, just by the words we use.

The English word 'emotion', for example, can be used to describe the full range of responses that people show when they are in the grip of an emotion such as anger, and it can also be used to include their conscious experience of feeling angry. Alternatively, it can be used in a more restricted sense to describe just the outward and visible signs of anger—the raised voice, the red face, the increased heart rate—without necessarily implying anything about the subjectively felt conscious experience of anger.

In the case of other humans, we happily blur the distinction between the different meanings of such words, and it usually does not matter that much. We assume that someone who looks angry also feels angry (unless they are a very good actor). But when it comes to other entities, it does begin to matter. 'An angry sea' is an obvious figure of speech. Seas aren't conscious, and they don't experience anything. But an angry wasp or an angry swarm of bees? An angry shark? An angry elephant? It matters very much which meaning of anger we intend here, specifically whether or not we mean to imply that the outward signs of anger are being given by something or someone who is consciously experiencing anger. By using the same word—'angry'—to mean yes, no, or maybe to conscious feelings, we inevitably sow the seeds of confusion.

The real hazard is not in fact the ambiguity of the words themselves. We can deal with ambiguity if we all realize it exists. If someone says that they found it 'funny' that the furniture in their room had been rearranged, we are in no danger of suddenly flipping from trying to help them solve a mystery to thinking there was something comic about the relative positions of chairs and sofas. The ambiguity of the words we use to describe possible states of consciousness, however, is more serious because not everyone realizes that there could be more than one meaning. Non-scientists are unlikely to ask 'Do you mean thought-with-consciousness or thought-without-consciousness?' because for most people, thinking means conscious thinking, emotion means conscious emotion, and mental events mean conscious mental events. The idea of thinking without consciousness just wouldn't occur to most people as a possible alternative meaning. It wouldn't even make sense because by definition thinking, to them, means conscious thinking. Such people might be genuinely surprised to discover that scientists and philosophers quite happily entertain the idea that at least some thoughts (such as what they call lower-order thoughts) can happen quite without consciousness.[5]

[5] Dennett (1991); Rosenthal (1993)

Unfortunately, language itself (at least the English language) often makes it difficult to be consistent in what we mean because it does not have enough words to make the distinctions we want to make. The field of pain research is a notable exception and shows how much clearer things can be when there are enough words to express different meanings. The physiologist Charles Sherrington was so concerned about the ambiguity of the word 'pain' that he went to the lengths of inventing a completely new word—'nocioception'—to distinguish the outward and visible signs of responding to injury from the conscious sensation of pain. By 'nocioception', he meant all the mechanisms—the surface receptors, the neural pathways, and the brain areas—that contribute to the way humans and other animals respond to bodily damage.[6] Nocioception may or may not be accompanied by pain. Pain, on the other hand, very definitely does mean the unpleasant conscious experiences that often accompany injury and is therefore, of course, much more difficult to study. Having these two separate words—pain and nocioception—means that scientists can make it clear whether, on any given occasion, they are just describing physiological and behavioural responses to injury or whether they have made the step into the altogether more uncertain territory of conscious experiences. Most people outside science, however, would not know what 'nocioception' means, so discussions about animal pain are all too frequently conducted with the one word—'pain' being used with two different meanings—one being neutral about whether consciousness is involved and the other declaring that it definitely is.

For most emotions and feelings, there are no technical words available to allow us to even try to say what we mean, so everybody, including both scientists and non-scientists, is in constant danger of using words with double meanings. Ambiguous words such as emotion, anger, cognition, or motivation positively encourage careless thinking. They allow us to get away with not defining

[6] Sherrington (1900)

what we mean or, what is much worse, to change what we mean depending on who we are talking to or what point we want to make. This is why I call deliberately using such ambiguous words flirting with consciousness. Someone might use a word such as 'anger' or 'fear' about an animal, knowing that the person they are talking to will almost certainly take this to mean that the animal consciously feels anger or consciously feels fear. Using such words makes the speaker sound as though they are working on some really profound issues to do with animal minds. But then, if they are challenged about what evidence they have that an animal really is consciously experiencing anger, they can always claim that, of course, they were only using these words in the restricted sense of what can be observed. They can reaffirm their scientific credentials by reiterating that we cannot know for certain whether animals are having conscious experiences or not. Like an accomplished flirt, they hold out a promise, but when things get serious, they can always claim that that was not really what they meant in the first place.

The result is, as I am sure you can see, a recipe for confused thinking. By using these ambiguous words, we don't just confuse other people. We confuse ourselves. Other scientists successfully use words that may have both a technical and an everyday meaning, but a physicist is unlikely to become confused about the difference between the 'charm' of a particle and the 'charm' of a politician. With words that may or may not imply conscious awareness, on the other hand, it is far more difficult to keep the meanings separate. We may not even be sure ourselves whether we are flirting or serious, and the real danger is that we may start off using a word in the scientific sense and then, if we use it enough times, slip seamlessly into the everyday sense. Consciousness can begin by being the hardest of all hard problems and end up being an assumption we make without thinking what we are doing, just because it has hitched a ride on the back of an everyday word.

As an illustration of the confusion that can be caused by making one word do the work that should really be carried out by two, take two recent papers in the same journal, *Animal Sentience*, that have subtle but crucial differences in their use of the word 'cognitive'. Miguel Segundo-Ortin and Paco Calvo, in a paper titled 'Plant Sentience' (2023), say clearly 'even if plants have surprising *cognitive abilities*, it remains debatable whether this is evidence that they *feel*', thus clearly making a distinction between cognitive-with-conscious feeling and cognitive-without-conscious feeling. Stevan Harnad, on the other hand, argues that 'Felt states can also include "cognitive" states, like what it feels like to recognize someone, to recall something, to want something, to know how to get somewhere, to understand a sentence, or to think something'.[7] He clearly sees cognitive states as a subset of felt states that do involve consciously feeling something. The stage is set for some major misunderstandings just by differences in the way a particular word, in this case 'cognitive', is being used.

The best way to avoid being seduced by words is to make—and to keep making—a clear distinction between complex information processing (memory, perception, decision-making, and so on) and the conscious feelings that may or may not accompany them. This takes considerable effort, as the double meaning of some words leaves us in constant danger of repeatedly blurring a distinction we thought we had already made, sometimes only a few moments before. It is so easy to be lulled into thinking that 'memory' always means consciously recalling the past, that 'perception' always means consciously seeing the world, and that 'decision-making' always means consciously agonizing over a choice between two alternatives. But that clearly is not the case. Our computers and iPhones are capable of doing many of the same things that we do. They store 'memories' and 'recognize' voices, but that does not mean that they have conscious experiences in the way that we do when we use the same words to describe our own capabilities. That is why it is vital to maintain this distinction

[7] Harnad (2023) p.1

if we are ever to successfully arrive at an answer to the question of which other animals are conscious. We know that animals can do clever things and can learn and do all sorts of 'information processing'. But we have to make sure we are focussing on the real problem—consciousness itself—if we wish to arrive at an answer to the question of which other animals have it.

It takes an effort to make meanings clear because it is tedious to have to keep saying things like 'fear-with-conscious-awareness of being afraid' rather than just 'fear'.[8] It also sounds pedantic, but at least it gives any subsequent discussion a chance of being a genuinely constructive exchange of ideas rather than two people talking past each other because both misunderstand what the other is saying or because one is serious and the other is just flirting.

In this book, our intentions are serious. We are going to be talking about the real thing—emotions and thoughts with consciousness—phenomenal consciousness if you like to use the term. So, although we begin our study of consciousness equipped with only a rough, working definition of consciousness as awareness or feeling something, our efforts are going to be greatly helped if we try to avoid the possible confusion caused by the ambiguous words that people use to describe what it is that they are aware of or what it is they feel. Since so many words commonly used to describe conscious states—such as fear, anger, stress, pleasure—can be used both to mean that consciousness is present and to mean consciousness might or might not be present, that is not going to be easy. However, as we shall see, the effort will be worth it. In the next chapter, we will see how technology has now made it possible not just to make this distinction but to cross and recross the boundary between the conscious and the unconscious mind.

[8] LeDoux (2014)

4
Other minds

In 1974, the philosopher Thomas Nagel published a paper with the unforgettable title *What Is It Like to Be a Bat?* in which he came to the disappointing conclusion that it is quite impossible for us ever to know.[1] Nagel called bats aliens. He argued that because they live in what is to us a strange world of sound made up of listening to echoes bouncing off objects in their environment, we humans can never know what a bat's world could possibly be like from the point of view of the bat. Human imagination is not enough for us to enter the subjective worlds of another species, he believed, because understanding what it is like to be a bat cannot be achieved by simply trying to imagine what we would feel if we had membranous wings, hung upside down all day and navigated using sonar. These are imaginable things, within human experience. Humans fly in gliders, and blind humans can learn to detect objects from the echoes that bounce back from them. The real issue for Nagel was not what we, as humans, might experience if we imagined we were a bat, but what a *bat* experiences as a bat, and we could only know that if we actually were a bat. And if we were a bat, then we would then not be us. We cannot simultaneously be a bat and be a human because we cannot get away from the essentially private, subjective nature of conscious experiences that each of us, separately, has.

The bat example enabled Nagel to express in a very graphic way the widespread belief that there is something uniquely difficult about trying to study consciousness in other species, something that makes it elusive to the point of being near-impossible to

[1] Nagel (1974)

bring it within normal scientific investigation. But is it really that impossible? After all, we now know a great deal about bats so that although they are aliens, they are not too alien. They are mammals like us, and their brains, although very different from ours, evolved from the same primitive mammalian brains that ours did.[2] Furthermore, we understand the physics of echolocation and know how it is possible to send out pulses of sound and then to use the returning echoes to build a sound landscape. We have even identified specialized nerve cells in the bat brain that are sensitive to the delays and distortions in these echoes so that it is no longer a mystery how a bat can catch food and avoid obstacles, all in complete darkness.[3] We might think, then, that this new knowledge would now make it possible for us to know what it would be like to be a bat. Nagel insisted, however, that for all our knowledge of how bat brains work from the outside, we would never know for certain what it is like to *be* a bat. We could imagine what it is like, but our ideas would always be incomplete, and there would always be something missing.

What would be missing is conscious experience, the subjective, the 'insider' view of the bat's world. But surely there is a problem here. Nagel's argument about the inaccessibility of bat minds also applies to the minds of other humans.[4] We cannot *be* another person any more than we can *be* a bat. We cannot know for certain what another human is consciously experiencing, a problem that philosophers call 'the problem of other minds'. So how is it that Nagel can argue that it is completely impossible to know what it is like to be a bat, however much we know about bat habits and bat biology but perfectly possible to know what it is like to be another human being just by using a bit of imagination and extrapolating from our own experience? His answer is that bats are just so different, so alien that our imaginations would not be

[2] O'Leary et al. (2013)
[3] Jones and Teeling (2006); Wenstrup and Portfors (2011)
[4] Kim (2011)

able to cope with something so different from our own experience. However, he provides no evidence that our imaginations are all that reliable with other people either. When we imagine what other people might be experiencing, we could, for all we know, be fanaticizing or making things up. Nagel himself admits that there are some human experiences that are beyond him to imagine, such as what it would be like to be a human born deaf and blind. Clearly, then, the distinction between knowing what it is like to be a bat (and Nagel assumes that it *is* like something to be a bat) and knowing what it is like to be another human is one of degree, not an all-or-nothing distinction. Indeed, one way of looking at Nagel's paper is that his bat example is a quite brilliant way of expressing the general problem of other minds, the problem of knowing for certain what, if anything, is consciously experienced by anyone—human, bat, or sea slug—other than ourselves.

Whenever we try to study any mind other than our own, we come up against the same problem. All conscious experiences go on inside just one head, and they are private. They cannot, by definition, be checked by anyone else, and this violates the most basic rules of the way good science is supposed to proceed: the results produced by one person must be checkable by anyone else and described in enough detail that they can be independently replicated. If one person claims that putting two stable chemicals into a test tube produces a completely different new, explosive, substance, that result has to be repeated and tested by other people before it can be accepted as a genuine discovery. It is this public nature of scientific results, their confirmation or refutation by different observers, that makes the difference between 'anecdote' and objective scientific 'fact'.

This private, subjective nature of consciousness was once taken to show that it is destined to remain forever in the realm of unverifiable anecdote. It led the psychologist John B. Watson to declare emphatically: *States of consciousness, like to so-called phenomena of spiritualism, are not objectively verifiable and for that reason can never become data for science.*[5]

Watson's views and those of like-minded scientists such as B. F. Skinner became known as Behaviorism, a way of thinking that dominated the way scientists thought about consciousness—human as well as animal—for much of the twentieth century. Or rather, it didn't so much dominate their thinking as stifle it altogether. Under the influence of Behaviorism, consciousness found no place in science and it was simply not studied. The logic of Behaviorism was simple and apparently compelling. Consciousness could not be measured objectively, and it was pointless trying to use it as an explanation of anything because it generated no new predictions that were not shared with its complete opposite—a total lack of consciousness. In other words, any behaviour shown by an organism that could be interpreted as being guided by a conscious mind could equally well be seen as the operation of an unconscious entity just going through the motions. There would be no way of telling which version was correct.

To quote Watson again:

> *One can assume the presence or absence of consciousness anywhere in the phylogenetic scale without affecting the problems of behavior by one jot or tittle and without influencing in any way the mode of experimental attack upon them.*[6]

Watson claimed that there is no task, no feat of memory, no evidence of intelligence that is unique to a conscious organism and that could not be mimicked by one that is not conscious. The hypothesis that a given organism is conscious therefore makes no unique predictions and a hypothesis that makes no unique predictions cannot be falsified. In his view, it was therefore not a scientific hypothesis at all.

[5] Watson (1929)
[6] Watson (1913)

Although Behaviorism now seems a rather extreme view to take, it does address an issue that, even now, will not completely go away. All the examples of consciousness we can think of—pain, seeing red, remembering a holiday, and so on—have in common that they appear to be restricted to an 'insider view' of the world. They all give us our own, private version of events that somehow excludes other people. It is as though you have been given privileged entry to an event where everyone else has to stay outside. These other people can see and hear something of what is going on, but you, from your unique position on the inside, have your own version and, because there was only one entry ticket, you are the only witness. You can describe to the people outside what is going on, but no one else knows at first-hand what you consciously see from your unique insider viewpoint. The private, subjective experience of the redness of your red, the hurt of your pain, the anxiety of your worry are yours and yours alone. No one can check them or say that you are wrong in the way that they can check that, say, malaria is caused by a parasite and spread by mosquitoes.

Conscious experiences thus appear to be place-dependent. They are totally different depending on the place where they are experienced. From inside one particular skull you could experience intense pain, but someone inside another skull might experience nothing at all. Since no one can be inside more than one skull at a time; we appear to be locked within our own skulls, and for this reason, consciousness seems destined to remain forever private and subjective—known and only knowable by one person. The problem of other minds looms as large as ever.

It is now realized, however, that although this private, place-dependent nature of consciousness certainly makes it difficult to study, it does not make it uniquely difficult, at least in other humans. There are many other things that are also dependent on the place where an observer is or the time at which observations are made, and yet people have found ways of studying them. None of us currently alive were present at the beginning of the universe or when the dinosaurs lived or when the Isthmus of Panama closed the connection between the Pacific and Atlantic Oceans. If you were in the UK, you did not actually experience the Turkish-Syrian earthquake of 2023 or see what was going on deep underground. We cannot *be* in the past or know in a direct sense what it was like to be in a different time or place. But we can put together very convincing indirect evidence (residual radiation, fossils, distribution of fauna, seismic records, first-hand accounts, etc.) that is public and open to skeptical scrutiny like any other scientific data. No one claims that physicists, geologists, or palaeontologists are unscientific just because they have to work with indirect evidence and were never actually present at events that took place in the past or on another continent or deep underground. It is their skill to make the best of fragmentary, indirect but nevertheless fully scientific evidence.

Consciousness is the same. We cannot *be* another person or experience exactly what they experience when they look at a sunset, but we do have a source of indirect, consistent, and very compelling evidence. What has changed since Watson's day is not what constitutes good science. Science still sticks (or should, anyway) to the same standards of being objective and concerned with publicly verifiable phenomena. What has changed is the view that conscious experiences are forever condemned to being unverifiable. They have now, with the help of new technology, been brought into science and given a new respectability. Within human neuroscience, there has been a realization that what people say about their subjective experiences actually constitutes reliable evidence about what they are privately experiencing. In direct contradiction to the Behaviorist view that what people say is just behaviour, verbal reports are now seen to be an acceptable account of what people are subjectively experiencing. In other words, introspection—once despised as non-science—is now the source of valid scientific data. Stanislas Dehaene, a pioneer in the study of human consciousness, put it this way:

> *Subjective reports are the key phenomena that a cognitive neuroscience of consciousness purports to study. As such, they constitute primary data that need to be measured and recorded along with other psychophysiological observations.*[7]

Note that when Dehaene describes what people say they are experiencing as 'primary data' about consciousness, he is not claiming that he can directly access the conscious experiences of other people in the sense of *being* another person or entering their private conscious world. By 'primary data', he means that what people say about their conscious experiences is valid evidence about what they are actually consciously experiencing. '*Subjective reports can and should be trusted*', he emphasizes in italics[8] and, what is more, he argues there is good evidence that they should be so trusted.

For example, from a wide variety of different experiments, it has been found that if people are shown a visual image that lasts for only a very short time (specifically, less than about 30 ms, which is 30 thousandths of a second), they report that they have not seen anything at all. But if that same visual image is presented for only slightly longer (50–60 ms), then most people report that they have, consciously, seen it.[9] The same effect is obtained in different countries and with completely different groups of people, so there is no way in which such a consistent result could be obtained if people were lying, mistaken, or just making things up. Conscious perception really does seem to have an objectively measurable threshold: a visual stimulus has to be seen by most people for at least 60 ms for it to enter consciousness.

The objective value of subjective reports is also demonstrated by studies in which what people say is found to correlate very closely with changes in objectively observable activity in their brains as measured by various imaging techniques. Functional

7 Dehaene and Naccache (2001) p.3
8 Dehaene (2014) p.42
9 Dehaene (2014)

magnetic resonance imaging (fMRI) is a completely non-invasive way of seeing—in real time—changes in which parts of a living brain are most active.[10] When a brain area is involved in processing information, the constituent cells start to use more oxygen. To meet this sudden need for more oxygen, the flow of blood to that area rapidly increases, which temporarily alters the amount of oxygenated blood surrounding the cells. Now, conveniently for brain imaging, oxygenated blood has different magnetic properties from deoxygenated blood, and so the changes in blood flow can be seen from outside the body by detecting changes in these magnetic properties, giving rise to what is called a BOLD (blood oxygenation-level dependent) signal. It is this BOLD signal that appears on a brain image map and shows which parts of the brain are active, or, rather, were active, a few seconds earlier because the BOLD signal reflects the results that brain activity has on blood oxygen levels, not the activity itself. After about 6 seconds, the blood returns to normal and the BOLD signal declines.[11]

fMRI and other brain imaging techniques have led to major discoveries about which parts of the brain are involved in different kinds of information processing.[12] For example, when people were asked to describe on a scale from +2 to −2 how pleasant or unpleasant they found the taste of a drink, there was a clear positive correlation between how pleasant they said they found it and the size of the BOLD signal in the ventral pre-frontal cortex of their brains. Sweet, creamy vanilla-flavoured drinks that they said were very pleasant resulted in a large increase in the size of the BOLD signal, while unsweetened watery drinks that the subjects said they found neutral or positively unpleasant showed much smaller or no change.[13] What people said so closely matched what their brain pictures were showing that observers were left in no doubt that what

[10] Logothetis et al. (2001)
[11] Devlin (2012)
[12] Raichie (2003)
[13] Grabenhorst et al. (2010)

they said they were consciously experiencing was deeply rooted in the objective, measurable reality of brain activity.

Being able to see into a living brain has also allowed scientists to take the next step and to ask whether there might be brain 'signatures' of consciousness—that is, activity that is specific to a conscious brain. The hope has been to identify changes in brain activity that could indicate the boundary between an unconscious and a conscious state, for example, as someone comes round from anaesthesia[14] or emerges from a coma.[15] There is now increasing evidence on such transitions[16] and there are also major changes in the electrical activity across the brain (known as the P300 wave) whenever a person becomes aware of something such as a word or an image.[17]

However, there are differences of opinion about what the observed brain changes really signify.[18] Although there are changes in the brain that are *correlated* with consciousness, they may not be indicators of consciousness itself. They could just be the prerequisites for consciousness, such as a sign that the brain is alert or that a memory is being retained.[19] To make matters even more complicated, it also turns out that virtually all the brain's regions can participate in both conscious and unconscious thought.[20] There are what have been described as consciousness 'hot zones' such as the parietal, temporal, and occipital lobes of the cortex,[21] but there is no consciousness centre. There is no one specific area of the brain where consciousness happens. There are no specialized nerve cells dedicated to producing conscious experiences. Consciousness has not been located, even in our own brains.

[14] Sanders et al. (2012); Mashour and Alkire (2013), Nilsen et al. (2022)
[15] Sitt et al. (2014)
[16] Cofre et al. (2020); Barttfeld et al. (2015); Mashour (2024)
[17] Dehaene and Changeux (2011); King et al. (2014)
[18] Farisco and Changeux (2023); Luppi et al. (2024)
[19] Havlik et al. (2017)
[20] Dehaene (2014)
[21] Koch et al. (2016)

This is, of course, somewhat disappointing in our quest for consciousness in other species. It would be very helpful if we knew more about what made our own brains conscious because that would give us a better idea of what to look for in other species. If it had turned out that a particular brain structure were associated with being conscious or not conscious in us, we could have asked if a similar structure were present in tree-shews, sharks, or octopuses and used this as evidence about whether they might be conscious too. Unfortunately, there is no such easy path to follow. The study of human consciousness has made major advances, but there is still much we do not understand.

However, the point of this chapter is not to dwell on what we still don't know about consciousness in humans. The point is to show that the discussions now taking place about it are based on objective scientific evidence. Over twenty-two different theories about human consciousness are currently under discussion,[22] and these theories will be judged ultimately like any other scientific theory—on whether they make predictions that turn out to fit the real world and can be confirmed by other scientists.

In other words, data on human consciousness are now regarded as publicly verifiable. Consciousness remains private in the sense of being subjective and experienced in a particular way by just one person, but the data on consciousness are not unknowable by others. Verbal reports, backed up by images of brain activity, careful control of the stimuli presented, and repeated on many different people, have become the gold standard for studying conscious experiences. They bridge the gap between observable behaviour and private consciousness, between the objective and the subjective.

Or at least they do for conscious events experienced and reported on by human beings. Human consciousness has become scientifically respectable because what people say they are

[22] Bayne et al. (2024); Mudrick et al. (2025)

experiencing has been accepted as valid, reproducible data—in fact, the crucial data about private, subjective events. That still leaves the problem of what data we can use when there are no words to provide a running commentary and no language to provide a hotline to the consciousness of another being. Obtaining comparable evidence about possible conscious experiences in animals without language might, therefore, seem as far away as ever, but that would be to underestimate the sheer ingenuity of scientists who have been determined to find consciousness even when there are no words to get us there. We turn next to the ways in which it has been claimed that consciousness in non-verbal animals has now also been brought in from the cold.

5

Other minds without words

In 1975, a large group of scientists from around the world converged on the beautiful Italian city of Parma for what turned out to be a momentous conference on animal behaviour. The 14th International Ethological Conference was, like its predecessors, an almost overwhelmingly intense few days in which all the latest ideas about animal behaviour were discussed among fellow enthusiasts from before breakfast until late at night. That much can be expected from any such meeting. But this particular conference was one that many of us will never forget because it quite literally marked a change in the direction that animal behaviour research would take.

Donald Griffin, whose groundbreaking book, *The Question of Animal Awareness* (1976), was to be published a year later, gave a talk in which he argued that people working on animal behaviour should start to include conscious experiences in the studies of their animals. To us now, that doesn't seem exactly a groundbreaking thing to say, but back then it was revolutionary. Based on careful and logical arguments, Griffin's talk was a battle cry against what he called the 'straitjacket' of the Behaviorist views we discussed in the last chapter. He urged scientists to be more broad-minded about animal consciousness and to give it its rightful place in mainstream science. The fact that this now just seems like common sense is testimony to the completeness of the overthrow of Behaviorism that Griffin pioneered.

The reason for starting this chapter with an outdated idea and its downfall is, however, to emphasize once again how much has changed. The last fifty years have been a period of immensely fruitful research into animal consciousness so that there are now thousands of books, journals, and research projects, all assuming that animals have minds and are conscious. However, the central claim of Behaviorism—that the concept of consciousness is unscientific because it makes no unique predictions—has still not completely gone away. It hangs around in the background as an unwanted ghost from the past, constantly reminding us that we cannot definitively distinguish between an animal that has conscious experiences from one that is behaving in exactly the same way but just behaving 'as if' it had them. We cannot escape from the fact that without verbal reports to give us the primary data about consciousness that we have with other people, we always come up against the same problem of how to reliably recognize consciousness when the animals in question cannot tell us in words what they are experiencing or even that they are experiencing anything at all.

It is important to be crystal clear about exactly why this remains such a problem. In the last chapter, we saw that what people say is now accepted as valid data for what they are consciously experiencing, and this can be backed up by the picture of what neuroimaging tells us is going on in their brains. People can provide a running commentary on their inner conscious lives, and this has put the study of human consciousness onto a sound scientific basis. Verbal reports have provided a way of making what was once considered inaccessible and private into objective data that can be publicly verified. Unfortunately, this only serves to highlight the difficulties of studying consciousness when there are no words to provide the running commentary. The study of animal consciousness lacks this key element—language—that has turned the study of human consciousness from subjective introspection into a proper science.

There are two different reasons why language is so crucial to the study of human consciousness. The first reason is the one we have already discussed—that language enables people to describe their conscious experiences and so to provide the primary data about those conscious experiences themselves. The second reason is that language also provides evidence of a particular kind of ability—the ability to reflect on thoughts and feelings. If someone says, 'I have a headache that feels as though my head is exploding', we are simultaneously receiving two separate kinds of information. One is that the person is in pain. The other is that they have a brain complex enough to reflect on what that pain is like and to describe it to us. Words organized into a complex language thus give us two kinds of information about other humans and that means that when there are no words, as happens with other animals, we are missing not one but two kinds of information.

The first kind of information we are missing would not seem to leave a particularly serious gap. The absence of words in animals could be seen as no more than the inconvenience of not having access to an important, although not vital, source of data. It would be like having to study evolution without access to the fossil record. Just as evolution can, if necessary, be studied without fossils using alternative sources of evidence such as molecular phylogenies and the behaviour of living species, so the study of animal consciousness might similarly focus on the necessity of making do with available alternatives, primarily non-verbal behaviour. The absence of words then just becomes a challenge to be met rather than an insurmountable barrier.

On the face of it, words would not seem to be essential to understanding conscious experiences at all, even in other humans. If someone whose language we did not speak was bleeding from an open wound, crying uncontrollably, and writhing on the ground, we would have no hesitation in saying that they were consciously experiencing pain. We would not need to wait for a translation of their 'verbal reports' into our own language to reach this conclusion. As many non-human animals have very similar behaviour to us when they are injured—limping, calling out, and so on—it would seem but a short step to conclude that they, too, feel pain in much the same way that we do. What would seem to matter most, then, is behaviour. Words seem to be secondary—the way we fill in details about what sort of pain someone has, where it hurts most, or how they came to be injured in the first place. With animals that have no words, we still have the primary evidence of their behaviour—their gestures, sounds, and body language—that Charles Darwin called 'The Expression of the Emotions'.[1] Darwin himself certainly believed that these provided powerful evidence of what animals were consciously feeling. If we agree with Darwin, then we would conclude that our main task in answering the question of which animals are conscious is in finding the right outward signs of consciousness—that is, behaviour and physiological responses—that we can use as substitutes for words. On this view, other animals are like people whose language we do not understand, so it would seem that we just need to take a bit of extra trouble to learn their language. Perhaps we just need to refer to a horse dictionary or a dog app that would allow us to translate—like Dr. Doolittle—directly from their language into ours. Can we then access animal consciousness through what animals do? Can body language do for animals what verbal reports do for humans and give us scientifically valid primary data on animal consciousness?

Unfortunately, the situation turns out to be more complex than this. There is the second kind of information that we miss out on when there are no words available. Without words, we lack the information that we are dealing with a brain complex enough to put thoughts into language and to describe its conscious experiences. An animal may be conscious, but it cannot tell us so. In at least one important respect, its brain is, therefore, not like a human brain.

This raises the question of how 'like me' an animal must be physiologically and anatomically for it to be 'like me' in having conscious experiences. With other humans, the same question can be raised because I can never know for sure that you consciously experience what I experience when I see red, but the similarities of our eyes

[1] Darwin (1872)

and brain make it a reasonable assumption. This is the 'problem of other minds' we discussed in Chapter 4, but it is not generally much of a problem in our everyday interactions with other people. The leap of faith I have to make to reach the conclusion that you are like me in having conscious experiences is a small one, because we are so anatomically similar. It is made even smaller by the fact that we can tell each other what we are seeing and even describe some of the inevitable differences between us. For all intents and purposes in everyday life, I can assume that you are like me in being conscious not just when you say you are but *because* you say you are.

With other species, however, the analogy is less close and the leap of faith needed to make it is correspondingly much larger. Our sense organs and brains are different, often massively so,[2] leading Nagel[3] and more recently Roige[4] to go so far as to argue that it is actually impossible to understand consciousness in animals because they are just too *unlike* us.

On the other hand, there are many other people who believe that these anatomical differences are relatively unimportant and we should be more impressed by the similarities, particularly when animals show cognitive skills that we know in ourselves are associated with being conscious, such as certain kinds of learning,[5] planning for the future,[6] understanding what is going on in another's mind,[7] or recognizing self in a mirror.[8]

Such abilities, it has been argued, are indicative of a conscious mind. One or two of these abilities on their own might not be enough to justify such a conclusion, but animals that show many of

[2] Yong (2023)
[3] Nagel (1974)
[4] Roige (2023)
[5] Ginsburg and Jablonka (2019)
[6] Cheke and Clayton (2012)
[7] Emery and Clayton (2001)
[8] Anderson and Gallup (2015)

them must be taken seriously as conscious entities.[9] The argument here is that if animals are 'like us' in being able to perform cognitive tasks that we know we do consciously, then the likelihood is that they are also 'like us' in doing those same tasks consciously. This argument has immense appeal. We are deeply impressed by animals behaving in a human-like way. We easily assume that their apparent intelligence must be an outward and visible sign of a conscious mind within. But how impressed should we be? This is where we need to ask some difficult questions.

As stressed at the beginning of this book, if all you want to do is to make an ethical decision as to how to treat a particular species of animal, then you may decide it is fine to base your decision on less-than-perfect evidence. You can assume that a clever animal must be a conscious animal. You can dismiss any criticisms as mere quibbles—unnecessary fussing over the remote possibility that they can do what we do but without consciousness. You can acknowledge that, of course, we can never know for certain whether they consciously experience anything at all but then say you are quite happy to conclude that, on balance, you think they are. The benefit of the doubt and the precautionary principle are powerful allies in favour of assuming that many animals, particularly the ones showing human-like cognitive abilities, are plausibly conscious.[10]

But as we saw in Chapter 2, this book is not just about looking for evidence for animal consciousness that is good enough to conclude that animals deserve the protection of the law or for you to decide that they need to be treated in certain ways. It is also about the altogether more demanding quest for consciousness itself. It is about deciding whether the evidence really shows what it is claimed to show and that what looks like consciously driven

[9] Braithwaite (2010); Broom (2014); Barron and Klein (2016); Birch et al. (2020); Mallat and Feinberg (2021); Mason and Lavery (2022); Dung (2022); Crump et al (2023)

[10] Birch (2017)

behaviour actually does have a conscious mind behind it. It may be entirely 'plausible' that insects, crabs, or plants have conscious minds, but how plausible is that? Possible, plausible, or very likely? To find answers, to find where the limits of sentience really are, we must be sure of what the evidence is really telling us and not be afraid to question it. So, how should we approach a claim that a certain kind of animal is conscious because it shows a particular behaviour or is able to perform a special kind of task? What tests should we apply to the evidence?

There is one key question we can ask ourselves to avoid being overly impressed by seemingly convincing evidence when we see an animal behaving in a particularly clever or human-like way. It is certainly not the only question we should be asking, but asking it has the salutary effect of making us stop and think. It provides pause for thought, time to consider a range of possible alternatives. Asking it is a habit well worth getting into.

The question is: *How easily could a computer do that?* This question does not imply that animals are just like computers or that they are 'nothing but' little machines. On the contrary, it is simply a useful part of a toolkit for testing out whether or not consciousness is genuinely present. As we saw in the last chapter, our quest to discover who is conscious would be greatly helped if we had a clear signature of consciousness—something that tells what to look for when we consider the variety of non-human animals that might be sentient. But a reliable signature must not just be an indication of consciousness present. It must also be proof against giving false positives and ascribing consciousness where none exists. Asking *How easily could a computer do that?* helps to eliminate these false positive ideas of what might indicate consciousness because people—all of us, that is—are notoriously easy to fool with the apparent cleverness of quite simple pieces of programming.

The question *How easily could a computer do that?* is thus not an attempt to deny that animals of many species may be conscious, but a way of turning the spotlight on the ones that are most likely to be.

Nor does the question rule out the possibility that one day, computers themselves may be conscious, as we will see in Chapter 7. By asking whether an apparently complex task could be easily done by a few lines of code, we have a quick and easy sanity check on our evidence. It doesn't rule out consciousness. But it helps to show how plausible the evidence for it actually is.

To illustrate how necessary such sanity checks are, we can start with a classic story of humans being completely fooled by a computer. ELIZA DOCTOR was a computer program developed over fifty years ago by Joseph Weizenbaum[11] at MIT to simulate the kind of conversation that might go on between a psychiatrist and a patient. Obviously, the computer could not be made to look remotely like a doctor, so the conversation had to involve a keyboard and a screen. The conversation might involve the 'patient' typing something like 'I never got on with my mother' to which DOCTOR would reply 'Tell me more about your mother'. This was remarkably effective, and many people who engaged with DOCTOR were quite convinced that the computer had human-like feelings and really understood them. In fact, all that DOCTOR had was a list of keywords to look out for, such as mother, father, angry, and so on, and a list of stock phrases to reply with, such as 'Tell me more about your … …' or 'What makes you feel so … …?' in which it would insert the keyword the patient had just used. If there were no keywords present in what the patient had just typed, DOCTOR would choose one of its list of standard bland phrases to keep the conversation going.

What was so interesting about ELIZA was how hard it was to convince most people that they were *not* talking to a conscious, thinking, feeling human being even though all ELIZA was doing was matching a relatively short list of keywords with another short list of phrases to respond with. An ultra-simple set of instructions was all it took to make many people feel that they 'just knew' it was a

[11] Weizenbaum (1966)

real person that was responding to them. There is even a story that Wiezenbaum's own secretary was one of those who always believed there was a human being in there somewhere. This illustrates the danger of being satisfied with assuming that because an entity has some of the attributes of a conscious being, it must have all of them, including being conscious. Intuition can be misleading and shows that what may look like the empathetic outpourings of a complex, conscious being can be easily mimicked by a few lines of code.

We, now, might imagine that we would not be so easily deceived. ELIZA DOCTOR was released into a world completely unused to computers, and so we might feel it is not at all surprising that people back then could be deceived by such a simple program. But if you think you are beyond being taken in by clever coding, I can assure you that we've all seen nothing yet. Artificial intelligence (AI) is already making it almost impossible to tell whether we are talking to a real human or whether an essay has been written by a real student or a machine.[12] Programs such as Large Language Models[13] can give a much more convincing imitation of human language than a primitive Chatbot like ELIZA because they have the capacity to learn from how real humans use language. Large Language Models learn what words are likely to follow from which other words when people speak or write and are then able to reproduce statistically similar strings of words themselves. Their output can be such a convincing mimic of human language because the models are based on so many examples. The whole of the internet is their training ground. Although the models may sometimes get things wrong and betray their non-human nature, they will soon learn to be even better at mimicking what we say because they will have even more examples of how we say it.

AI challenges, and will continue to challenge, all our preconceptions about what computers can do, including having conscious

[12] Kasneci et al. (2023)
[13] Chang et al. (2024)

experiences. But for now, as far as we know, no machine is conscious and yet machine intelligence is increasingly able to mimic human behaviour. So more than ever, we need to ask what a mindless machine could do before we jump to the conclusion that, with other species, we are necessarily dealing with the workings of a conscious mind. This is not—to repeat, not—because animals are mindless machines but because of the importance of not basing our arguments on the tempting assumption that 'a computer could never do that'. If you don't realize how clever computers already are, you are more likely to erroneously conclude that an animal doing something remarkable or clever must be conscious too. That animal may well be conscious, but mere human-like behaviour is not enough to conclude that it is.

Let us now see what happens when we ask the question *How easily could a computer do that*? of a number of criteria that have been widely cited as showing consciousness in animals, not to discredit the evidence or to argue that animals behave unconsciously, but merely to give pause for thought before consciousness is unquestioningly accepted. The list is not intended to be exhaustive but just to give an idea of how fragile some of the claims for animal consciousness turn out to be.

Integrating information: How easily could a computer do that?

A theory of human consciousness that has gathered widespread support over recent years is the 'global workspace' theory. Originally put forward by Bernard Baars[14] and subsequently modified by various people,[15] global workspace theory is the idea that consciousness comes about when information is broadcast across the brain, allowing different regions to communicate with each other.

[14] Baars (1988, 2002)
[15] Dehaene (2014); Barron and Klein (2016); Seth and Bayne (2022)

This global pooling of information is then held to lead to a conscious state of mind in which flexible decisions can be made based on all the available evidence—external stimuli, internal stimuli, memories of past events, and so on. Consciousness is thus seen as resulting from a kind of brain teamwork or cooperation between different brain areas. Unconscious processing, by contrast, is held to be confined to small, localized parts of the brain, resulting in fixed, less flexible decisions.

There is certainly evidence that when people report conscious experiences, many different parts of their brains are active, particularly the pre-frontal cortex.[16] However, widespread brain activity does not seem to be a defining characteristic of consciousness since unconscious activities can also involve many different brain areas, even in humans. For example, jumping to catch a ball involves the highly complex integration of different sensory and motor parts of the brain, but skilled ball-catchers are not conscious of how they manage to put their hands in exactly the right place at exactly the right time.[17] In any case, the global workspace theory does not specify how much integration is needed to produce consciousness, that is, just how 'global' the global workspace has to be. It just says that as information processing becomes more widespread, at some point consciousness happens, but not where this point is or, more importantly, what sort of information processing makes the difference between what is done unconsciously and what is done consciously.[18]

Modified versions of the global workspace theory have attempted to address this vagueness problem by being more precise about exactly what sort of information processing might be associated with consciousness. Barron and Klein,[19] for example, argue that the detailed control over movement shown by flying

[16] Dehaene (2014)
[17] Reed et al. (2010)
[18] Rolls (2014)
[19] Barron and Klein (2016)

insects, such as flies and wasps, involves integration of sensory, motor, and homeostatic information from many different parts of the brain and is in this sense 'global'. On this basis, they then argue that these insects can consciously feel pain, a conclusion that is now supported by an increasing number of other authors,[20] but without a convincing answer to the most serious criticism of global workspace theory itself—the lack of clarity as to what these interconnections are supposed to be doing to support conscious experiences. So when Adamo[21] disagreed and argued that the insect brain is not interconnected enough to produce consciousness, there is no real way of deciding who is right because no one can say what the connections between different parts of the brain have to be doing to make the jump into consciousness.

Driverless cars are quite enough to show that computers are perfectly capable of integrating vast amounts of information without having to be conscious. Driverless (autonomous) cars receive inputs from a variety of different sensors (cameras, radar, GPS, etc.) and use complex algorithms and machine learning to navigate their way down busy streets. They learn about their environment and create maps that are constantly updated. They use their experience to make split-second decisions about whether an approaching car is a danger or is about to pass harmlessly in the correct lane. In other words, driverless cars show all the signs of using a very global workspace indeed, and so unless we want to attribute consciousness to driverless cars, applying it to animals just because some parts of their brains are connected to other parts is not convincing. Global workspace theory would be much more credible and much easier to apply to non-human animals if it could specify exactly what functions all the broadcasting and integration of information had to perform to give rise to consciousness. As it

[20] Huis (2021); Crump et al. (2023)
[21] Adamo (2016, 2019)

is, integrating information from different sources is exactly what mindless computers are very good at.

Motivational trade-offs: How easily could a computer do that?

Trading off one motivation against another is also an ability that has been claimed to indicate that a conscious mind might be at work.[22] A motivational trade-off occurs when an animal is motivated to do two incompatible behaviours at the same time and has to make a decision about which one to do. For example, an animal might find itself in a conflict between whether to approach a river to drink because it is thirsty or whether to stay away from the river because of fear of being attacked by crocodiles. Such an animal would have to 'decide' whether it is worth enduring the negative consequences of possibly being attacked in order to achieve the positive benefits of having a much-needed drink. If it shows evidence of being able to weigh up the costs and benefits of doing one or the other of these options, then the implication is that it might be consciously deciding what to do and, specifically, capable of imagining what might happen if it chose one over the other.

Robert Elwood decided to put hermit crabs into a motivational conflict by delivering a small electric shock to their soft abdomen inside the mollusc shell in which they had taken up residence.[23] The crabs had therefore to decide whether to get out of the shell to avoid the shock or to stay inside their protective home. The results depended on whether the crabs were in a preferred or a less preferred shell. Elwood had previously established what shells the crabs preferred by giving them a free choice among a range of sizes and species and seeing what they chose. Crabs that were living in

[22] Crump et al. (2023)
[23] Elwood and Appel (2009)

a shell of their preferred species were less likely to get out of their shell when shocked than crabs that had been making do with a less preferred shell. The hermit crabs were thus resolving their motivational conflicts in a way that suggested they were able to trade off the costs and benefits of being shocked against those of having to abandon a more or less valuable home. Even this, however, does not necessarily mean that they were consciously deciding which option to take.[24]

'Deciding' is one of those potentially seductive words we discussed in Chapter 2. It can have more than one meaning. Despite its usual pairing with consciousness in humans, 'deciding' can in fact refer to the resolution of a conflict that is done either with or without consciousness.

Once again, here is something that it is easy for a computer to do. In fact, being able to decide between different courses of action is essential for any robot that has to function autonomously and without human guidance for any length of time. Even robot lawnmowers can be said to 'decide' when it is necessary to stop mowing and go to recharge their batteries, but imagine a robot sent to a distant planet that was too far away for instructions from Earth to reach it except with a huge time delay. The robot would have to make a large number of decisions for itself about whether and in what order to do its various tasks. If it had been designed to map the planet and to collect geological samples from various locations, it would need a way of prioritizing or trading off its different options. These decisions about which behaviour to do next could be programmed in a very flexible way so that a machine could constantly adjust its priorities over time, depending on how much charge its battery has left, how far away the charging point is, how much charge it is going to need to fulfil its next task, and so on. A system of motivational priorities based on trading off the effect of doing one task rather than another would be as fundamental to the effective working of

[24] Elwood (2022)

an autonomous machine as it is to animals. In both cases, effective management of time, resources, and potential threats is essential and, in the case of our robot, not a particularly difficult programming task. It is certainly not one that needs consciousness to carry it out.

Protecting the body from injury: How easily could a computer do that?

Most animals show evidence of nocioception, that is, the ability to detect and respond to dangers such as heat, chemical, or mechanical injury. Insects, for example, have a variety of different surface damage-detectors that connect to the brain and allow the insect to escape from or mitigate damage.[25] The crucial question, of course, is whether these basic nocioceptive responses are or are not accompanied by conscious experiences of pain.

The evidence that insects actually do feel pain is that their nervous systems can do much more than just detect a surface problem and initiate a fixed response like a car that automatically switches on its wipers when it detects water on its windscreen. The insect nervous system has a complex system of two-way communication between brain and surface receptors so that the brain can actually change the volume of messages coming in from the surface, turning it up when the message is urgent and turning it down when other behaviours have higher priority. This ability to modulate nocioception is also found in mammals, leading Matilda Gibbons and her team to argue that 'the presence of descending controls makes it at least plausible that insects have painful experiences'.[26]

[25] Gibbons (2024)
[26] Gibbons et al. (2022)

Bumble bees have also been shown to 'self-groom', that is, rub their antennae when these are touched by a heat probe,[27] much as we might rub an injured knee or elbow. This, too, has been suggested as evidence of consciously felt pain, but there is an important proviso to make here. Responding to bodily threats and needing to repair damage are not unique to humans or other animals but would almost certainly emerge as a feature of any truly autonomous entity such as the robot on a distant planet that we invoked above. Self-care would be a likely requirement for any entity having to function on its own for any length of time.

As of now, there are no truly autonomous robots, at least not in the strong sense that animals are autonomous. Animals can quite literally take care of themselves—find their own food, repair themselves, and reproduce, and no robot can yet do that. But the robots of the future—such as the ones we send to other planets—are likely to be developed to the point where they have to function without us to constantly look after them. Obviously, it is unlikely that these robots will be made of the same flesh and blood as we are, so their damage avoidance and repair systems will be different from ours. But metal and plastic get damaged and corroded, wheels fall off, and inlet fans get clogged with dust. Whatever they will be made of, the robots of the future will have the equivalent of our injuries that could potentially put them out of action, and they will have to take protective action if they are to keep working. They will therefore need surface-damage detectors to warn them when something is wrong and they will need to use this information to take appropriate action. In other words, they will need to have a fully functional nocioceptive system capable of determining where any damage has occurred and of deciding what avoidance or protective behaviour to initiate. It will be an interesting coding exercise, but it could be accomplished entirely without consciousness.

[27] Gibbons et al. (2024)

Self-medication: How easily could a computer do that?

Many species of animals that are ill or injured will voluntarily give themselves analgesics, that is, drugs that are known to relieve pain in humans.[28] For example, rats with arthritis will learn to give themselves doses of fentanyl, a powerful synthetic opioid analgesic drug, but if their arthritis is treated medically, their intake of fentanyl drops.[29]

The inference often drawn from such studies is that just as we humans take analgesics to relieve our conscious pain, so do animals that learn to self-medicate because they also want to relieve conscious pain. Animals that self-medicate and can then walk or behave more normally are, on this view, living proof that they once consciously experienced pain and now no longer do so. The similarity to our own behaviour when we take a tablet to relieve a headache is so striking that to say that other animals are not consciously feeling anything under these circumstances just seems perverse and even heartless. But the robot left to fend for itself on another planet might have to employ a very similar kind of behaviour without necessarily feeling anything at all.

The challenge for the designers of this robot is that they will have only an incomplete idea of what the environment their creation will encounter in the real world, particularly if that world is another planet. When the designers sit down to plan their robot, they will somehow have to anticipate a broad range of problems it might just possibly run into, such as the robot falling into a crevice, having a part broken off by a falling rock, or losing power. How would they design such a machine? How could they write its software to be so open-ended as to be able to cope with environments very different from anything the designers were familiar with and with events that they might never have thought of? They could do a

[28] Gutierez et al. (2011); Pham et al. (2010)
[29] Colpaert et al. (2001)

lot worse than use the way animals have evolved to cope with a strikingly similar problem.

Natural selection has come up with a great solution. Instead of going through life with an infinitely long list of fixed behaviours, animals have evolved a system of goals or desirable end-states plus the ability to learn how to achieve those goals. So, for example, instead of having it hardwired into them exactly what their food must look like and exactly how it should be eaten, most animals have a goal of, say, seeking out a sweet taste and then they learn by experience how to achieve that goal. This gives them great flexibility because by seeking out the reward of this sweet taste, they can exploit new food sources that none of their ancestors had ever encountered. By having a system of rewards (desirable outcomes) and punishments (outcomes to avoid), animals can then learn to cope with entirely new environments and new opportunities as well as being much better equipped to deal with unexpected hazards. Biologically, this is achieved by having genes that specify what they find rewarding or punishing (their goals), guiding them towards what is beneficial and away from what is harmful. But then they also have the capacity to learn from their own experience what to do to achieve those rewards or avoid those punishments.[30]

At this point, we must be careful to avoid yet another word trap. Words such as 'rewarding' and 'punishing' can be used to imply conscious experiences. If someone says that they found helping out at the village fête very rewarding, we might naturally take it to mean that they had consciously experienced pleasure. Similarly, if they said they talked about a punishing medical treatment, we might naturally take it that they had experienced something unpleasant. But rewards and punishments do not need to be conscious to be effective. Machines learn by algorithms, which can be nothing more than simple rules such as if behaviour A is followed by result

[30] Balleine and Dickinson (1998); Dickinson (2012); Rolls (2014); Pennartz et al. (2019)

B, increase the probability of doing A. The association is made, and the output is modified by nothing more than a systematic change in the value of a few variables. For a machine, a reward is nothing more than a result that increases the probability of an action being repeated and a punishment nothing more than a result that decreases that probability.

So, an effective way of building a robot that could operate on a distant planet without constant human intervention would be to give it the functional equivalent of the reward and punishment system we find in animals. This would give the robot the means of removing itself from a whole variety of situations in which it was damaged or likely to be damaged and the ability to learn not to repeat actions that led to that damage. The robot would need to be programmed in advance to recognize and avoid situations that threatened its viability wherever it found itself, such as being trapped and unable to move or exposed to very hot temperatures. These situations would be the machine equivalents of negative goals or 'punishments'. But exactly how the robot would respond to being punished could be left to its own ability to learn what most effectively diminished the negative state of punishment and increased the positive state of 'relief'.[31] The result would be escape strategies specifically adapted to the conditions on the new planet, even though these might be very different from those on Earth or anything that could have been anticipated by the original designers.

To be really effective, however, such a damage-avoidance system would also need a means of establishing whether what it was doing was making things better or worse. In other words, a system for escaping from danger also needs a way of monitoring progress or lack of it from the resulting action. If the action results in the punishment actually increasing, then something else should be tried. If you are in a hole, stop digging, but if what you are doing results

[31] Navratilova et al. (2013); Gerber et al. (2014); Porreca and Navratilova (2017)

in the punishment signal gradually getting less, that suggests the action is the correct one and should be continued or repeated. Cessation of aversive stimulation ('relief') is a powerful reward for us, and we take every opportunity, for example, through pain-relieving drugs, to achieve it.[32] The fact that animals, given the chance to give themselves the same drugs, also find them positively reinforcing should therefore be no more surprising than that they find food and warmth rewarding.

The important point here is that because relief (a 'things-are-getting-better' detector) is so integral to effective damage avoidance in any system, it can be expected to be built into an unconscious autonomous robot as part of making it work effectively. As much as any animal, the robot would need information about whether its actions were being effective or not. But this also carries with it the danger of a severe malfunction. In an extreme case, a robot might disable its own sensors because this would give it the reward of the (false) relief from a signal that it was no longer in danger. It would be doing the equivalent of us turning off a smoke alarm because we don't like the sound and ignoring the fire. Learning to find situations that reduce punishments and provide the reward of relief is therefore not something that is a special hallmark of consciousness. It is behaviour that we should not be surprised to find in any entity that was built to learn by using a goal-based system of rewards and punishments.

With a machine using such a goal-based system, it would be necessary to prevent it from achieving a state of false relief by pre-programming it with a special instruction forbidding it to disable its own damage sensors, rather in the way that we might specify that a smoke detector must never be switched off, however annoying its sound is. Otherwise, the autonomous robot might remain exposed to dangerously high temperatures or remain forever stuck in its damaged state, falsely acting as if nothing more needed to be done.

[32] Gerber et al. (2014); Porreca and Navratilova (2017)

Just having a built-in goal of noxious stimulation without a forbidden clause would be enough for it to give a very good imitation of 'self-medication'—the equivalent of what we do when we take enough painkiller to allow us to walk on a swollen ankle and then run the risk of causing more damage). As with previously suggested criteria for consciousness, learning to self-administer a painkiller could be indicative of a conscious mind, but then again it might not be. It could just be a by-product of any goal-directed system capable of modifying progress towards its goal—in this case, the goal of reducing the incoming signal that the body has been damaged. 'Relief' is a state brought about by the reduction in a negative stimulus and it may or may not be accompanied by consciousness.

Mind-reading: How easily could a computer do that?

The ability to understand the mind of another individual might seem the undoubted hallmark of a conscious mind or, rather, two conscious minds. Any animal that has a Theory of Mind[33] must surely have a mind of its own. To investigate this, we can turn to a fascinating experiment on scrub jays by Nathan Emery and Nicola Clayton.[34] Scrub jays are an American species of jay that store food items such as acorns by burying them in the ground when there are plenty around and then digging them up again when food becomes scarce. This is an excellent survival strategy, but it is also open to the danger of the food being stolen, particularly if another bird has seen where the food was hidden in the first place. Emery and Clayton reasoned that if scrub jays were aware of this danger and knew that they had been observed burying food by another bird, they would be more likely to move their food to a different place. Behaving differently depending on whether or not they had been observed by

[33] Premack and Woodruff (1978)
[34] Emery and Clayton (2001)

a potential thief might therefore indicate a conscious awareness of the problem of thieves and what might be going on in their minds.

Scrub jays turned out to adapt to life in the laboratory very well, even to the point of obligingly hiding food in exactly the places the experimenters wanted—namely, sand-filled ice-cube trays that served as cache sites. Emery and Clayton devised a situation in which their scrub jays were given the opportunity to hide food under two different conditions. In one condition, a jay was given food to hide while being observed by another bird through a clear plastic screen. Under the other condition, a jay would also be given the same food to hide with the same options for hiding it, but this time it could do it 'in private' where the plastic screen was covered with a towel and so no other birds were visible. All the subject birds hid their food in the ice trays in both conditions. After three hours, a subject bird would be allowed back to its caching tray, but this time there was also a second tray that looked different and was in a different location. The bird therefore had a choice between digging up and eating the food immediately, leaving the food where it was in the first tray or digging it up and re-burying it in the new second tray. Birds that had been observed hiding food by another bird in the first part of the experiment were much more likely to remove the food from the original tray and re-cache their food in the second tray, whereas those that had buried food in private tended to leave it where it had been originally hidden. The jays had thus clearly remembered whether their food storing had taken place in the presence of another bird three hours earlier and then, if it had, took action that would reduce the risk of the food being taken by the potential thief.

Emery and Clayton called this result a 'good candidate' for evidence that jays can attribute knowledge and intent to another bird—in other words, evidence that the jays are not only conscious but attribute consciousness to another being. They have a theory of mind. On this interpretation, the birds were consciously aware that another bird had seen them burying food, inferred that this observer bird intended to steal it and worked out what to do to stop this from happening. One conscious mind worked out what was going on inside another conscious mind. A simpler, non-conscious explanation was, however, subsequently provided by a computer simulation of the jays' behaviour using just a few lines of code.

Van de Vaart et al.[35] were able to mimic the behaviour of the food-caching jays by simply assuming that the presence of another bird during the food-hiding stage would result in an increase in a variable they called 'stress'. The level of stress would then determine the probability that a bird would dig up and re-bury food when given the chance a few hours later. Thus their computer algorithm simply involved specifying that a virtual bird that had originally cached food in private would have a low stress level throughout the experiment and would leave the food where it was originally hidden, whereas a virtual bird that had endured the stress of being watched by another bird during the first part of the experiment would have a much higher level of stress and therefore a greater tendency to move the food to a different hiding place. By adjusting the time for which stress levels were maintained and the threshold level of stress at which re-caching was triggered, the behaviour of the simulated virtual birds very closely matched that of the real birds, but without the need to invoke any theory of mind or any inkling of consciousness.

Jays may or may not be conscious and they may or may not know what is in the mind of another bird. Indeed, Thom and Clayton[36] subsequently argued that 'stress' could not account for the caching results. However, the point of showing that the jays' behaviour can be simulated with a simple behavioural rule based on the level of a single variable (whether it is called 'stress' or not) is not to show that this is the way real birds do it. The point of the

[35] Van de Vaart et al. (2012)
[36] Thom and Clayton (2013)

simulation—of asking whether a virtual bird running on a computer could do the same thing—is to widen the field of possible explanations and to introduce an alternative and simpler hypothesis that people might not have thought of. Anyone unfamiliar with computer simulations, for example, might simply not have realized how easy it would be to instruct a computer to mimic the behaviour of real birds hiding and then re-hiding food. They might easily have thought that the behaviour was so complex that a conscious mind was the only possible explanation.

At the risk of being boringly repetitive, I am going to make the same point I have made before because it is so important and so easily misunderstood. Asking how easily a computer could do what an animal does is not a means of showing that an animal is or is not conscious, but it does put more runners in the race. It stops us from being gullible. It expands the range of possible explanations to be eliminated and it prevents us from assuming that consciousness is the only possible explanation behind what an animal is doing. The next example will show even more clearly how important it can be to consider a range of possibilities, not just the one of consciousness.

Self-monitoring: How easily could a computer do that?

In humans, one of the capacities most closely associated with consciousness is the ability to think about our own thoughts—that is, to think about such things as our memories or plans for the future and to monitor them for their accuracy, feasibility, and so on. These are known as Higher Order Thoughts or HOTS.[37] For example, if you think it might be fun to meet up with a group of friends on a particular day, you can think about where you might meet and how easy it would be for different people to get there.

[37] Rosenthal (1993, 2005)

Then, in thinking about your planned day out, you might realize that the date you had chosen was not a good one because it clashed with something else you were doing. In other words, you could evaluate your own thoughts to see if they would be likely to produce a desirable course of action. You could even pinpoint which part of your original plan was not going to work (location fine, date impossible) and change it accordingly, all the while doing nothing except thinking and, specifically, thinking about what you are thinking.[38] Such HOTS are often considered to be the exclusive hallmark of a conscious mind. So considerable excitement greeted the publication of a paper in 2001 by Robert Hampton that claimed to show that rhesus monkeys were capable of doing this too.[39]

Hampton argued that monkeys can monitor and assess the strength of their own memories based on an experiment in which he showed that monkeys behaved differently, apparently depending on how well they thought they had remembered what they had seen some time earlier. The monkeys thus appeared to have HOTS in the sense of being able to monitor the strength of their own memories. If monkeys can really do this, this might be the best evidence yet of HOTS in a non-human animal and thus the workings of a conscious mind.

Hampton trained two rhesus monkeys on a memory task that involved them having to remember a visual image for long enough to pick out that image from a set of alternatives appearing a short time later. Each monkey would sit in front of a computer screen that would show the first image—say, of a chicken. The picture would then disappear and after a delay (initially just over 30 seconds), the monkey would be shown a second screen now with four images—the chicken and three new ones. The monkey's task was to touch the image that was the same as the one it had seen

[38] Rolls (2020)
[39] Hampton (2001)

before. If it remembered that it was the chicken it had seen earlier and touched the correct image, it would receive a reward in the form of a peanut, a treat much favoured by monkeys. If it touched the wrong image, it received nothing and the screen went completely blank for 15 seconds. This was the Memory-high-reward task: the sample image had to be remembered over time, and the reward was considerable, but there was a risk of getting no reward at all if the monkey got it wrong.

Simultaneously, the monkeys were trained on a second and much easier task in which they were also shown a sample image for the same amount of time, but this time when the second screen appeared, there was only one image to choose from—just the chicken, for example. All the monkey had to do this time was touch the screen. It didn't matter where it touched, it always received a reward. The catch was that in this simpler task, the reward was only a pellet of monkey chow, something much less favoured by monkeys than peanuts. This easier task we can call the No-memory-test-with-low-reward task because it did not require the monkey to remember anything about the sample image to get a guaranteed reward, but the resulting reward was much less valued by the monkeys.

Once the monkeys were reliably doing both the Memory and the No-memory tasks, Hampton introduced a twist. On some occasions, he, as the experimenter, would decide which task a monkey received. He could make them take either the Memory-high-reward task or the No-memory-low-reward task. On other occasions, he let the monkeys decide for themselves which task they wanted to do. What he found was that the monkeys did better when they themselves decided which task to go for than when he decided it for them. Specifically, the monkeys obtained more rewards on the Memory-high-reward task when it was their choice to do it. On these trials, they behaved as though they knew the strength of their own memories. If they had a strong memory of the sample image and were confident they could get the Memory-high-reward task right, they would go for the difficult, high reward option, resulting in a high degree of accuracy. On the other hand, if they had only a weak memory of the sample image, they would be uncertain that they could pick the correct image out of the four on the second screen and they would go for the safer No-memory-low-reward option that guaranteed them at least a low-value reward. They appeared to adjust their behaviour according to how well they had remembered the sample image.

This strategy was only possible, however, when they themselves could choose which task to do. When Hampton decided for them which task they were to do, he had no idea how well they had remembered the sample image on any given trial. So if he gave them no option but to do the Memory-high-reward task, the monkeys would be forced to choose among the four test images, regardless of how well they had remembered what the sample image was. On some of the trials, they would happen to remember well and be able to choose the correct image, but when they had not remembered very well, they would inevitably do badly and fail to get any reward at all. Hence their greater accuracy when they could choose which sort of trial to do.

This experiment can (and has been) be interpreted as showing that monkeys can think about the state of their own memories, monitor how well they can remember things, and that they are therefore having HOTS. For many people, this indicates consciousness. There is, however, another explanation, one that comes from asking how easy it would be for a computer to produce the same behaviour as the monkeys. The answer is that it would be very, very easy.

All a computer algorithm would need to do to replicate the monkey's behaviour is to represent the strength of the monkey's memory of the sample picture by a single variable, call it M. M takes on a certain value when the virtual monkey sees the first image. Perhaps some images are more potent or more memorable than others, so the initial value of M will vary. Then over time, M is set to decay like a human memory, at a rate that will also vary with all sorts of factors. When the time comes for the virtual monkey to choose whether to go for the Memory or the No-memory task, the algorithm states that if the value of M is above a certain threshold, the virtual monkey should choose the Memory task with a high chance of getting a valuable reward but that if M is below this threshold, it should opt for the No-memory task and so be sure of getting at least a low value reward. The rule is in principle no more complex than your car monitoring how much fuel or battery it has left and warning you when and only when this drops below some particular level. Obviously, our cars do not need HOTS to have effective monitoring of their internal state and to be able to tell us when they need fuel or to be plugged into a charging point. Similarly, our bodies constantly self-monitor for all sorts of variables, such as blood glucose levels, of which we are completely unaware, let alone have HOTS about. Conscious thoughts about thoughts are thus one form of self-monitoring but not all self-monitoring has to involve conscious thought. Once again, the fact that a computer can mimic the behaviour of an animal does not show one way or the other whether the real monkeys were consciously trying to work out how good their memories were. They may well have been. But the lesson we should take from this is that self-monitoring is very easy for computers. It is certainly not the prerogative—or the signature—of a conscious mind.

Self-recognition in a mirror: How easily could a computer do that?

The ability to recognize 'self' is yet another possible indicator of a conscious mind. In 1970, Gordon Gallup published a landmark test for whether a chimpanzee could realize that when it looked in a mirror, it was a reflection of itself that it was looking at.[40] He started by just giving chimpanzees mirrors and found that they were quite fascinated by the sight of their own reflections. He also noticed that, as they became more familiar with their mirrors, the chimpanzees started using them to examine parts of their bodies, such as their teeth, that they could not see directly. This suggested that they actually realized that it was themselves they were looking at in the mirror, not some other member of their own species, as happens, for example, when cats or birds fight themselves in a mirror. Then Gallup did a critical experiment. He lightly anaesthetized a number of chimpanzees and then put some bright red, odourless dye on their eyebrows and ears. His idea was to make a sudden change to the chimpanzees' appearance that they could not feel or smell and would only know about if they looked in the mirror. When his chimpanzees came round from the anaesthetic and were fully awake, he gave them a mirror and watched what happened. Typically, a chimpanzee would initially scrutinize its reflection but then something remarkable would happen. Most of the chimpanzees stared into the mirror and touched the patches of red dye on their faces, but, crucially, not the patches on the mirror chimp, but the patches on their own face. Gallup took this to mean that the chimpanzees had not only noticed that there was something different about their reflections but had also realized that this difference was caused by something that had changed about their own face. The chimpanzees thus realized that they were looking at themselves in a mirror. This ability to pass the 'mark test' as Gallup's method came to be known is now used as evidence that chimpanzees (and a few other animals) are self-aware, that is, they have a consciousness of self.[41] The mark test has been criticized and refined,[42] and there are doubts about which other animals have actually passed the mark test, but the claim that mirror recognition

[40] Gallup (1970)
[41] Anderson and Gallup (2015)
[42] Gallup and Anderson (2020); Kakrada and Colombo (2022)

is evidence of consciousness of self is widely believed.[43] If this is true, it would surely provide a major test of what a computer could or could not do.

As far as I am aware, no one has actually done a mark test on a robot and, in any case, it might not be a fair test for an entity not used to grooming itself or responding to visual changes to its surface since robots are still at the stage of relying on humans to fix and clean them. But there is already evidence that robots can learn to recognize their reflection in a mirror, or at least to respond in very distinctive ways to their mirror image. For example, TIAgo is a robot developed by the Technical University of Munich[44] that is able to distinguish between its own mirror reflection (which always moves its arms at the same time as it does) and an identical looking robot (which moves its arms at different times). If you think about it, this is not a particularly difficult thing to do. If you stand in front of a mirror, you will see a perfect correlation between your own movements and those of your reflection, whereas there will usually be very little, if any, correlation between your own movements and anyone else's. So, if all that is needed for mirror self-recognition is learning to detect a perfect correlation, then it is an easy task for a computer to learn. It would simply have to learn that mirrors are a distinctive type of object with the peculiar property of providing 100% correlation with self-produced movements. And if it could do this, it would be an easy step to learning the connection between the visual image of a moving limb and the stimulus of touch on a particular part of the robot's exterior. No consciousness would be needed, just the learning of how different stimuli correlate with each other and with the movements that are being made.

This of course raises the issue of exactly what it is that animals are doing when they learn to use mirrors and whether mirror

[43] Suddendorf and Butler (2013); Gallup and Anderson (2020)

[44] Technical University of Munich

self-recognition is really the evidence of consciousness of self as has been claimed. When they are first exposed to mirrors, all species (including humans) tend to start out by seeing their mirror image as another member of their own species, often attacking them as intruders. At one time, it was claimed that only a select few animals (chimpanzees, killer whales, Asian elephants, magpies) could pass the mirror test and eventually learn that the behaviour of their mirror image reflected what they themselves were doing.[45] However, it has more recently been shown that cleaner fish[46] and even ants[47] are capable of passing the mirror test, raising the question of exactly what cognitive abilities are needed to 'pass the mirror test'.

Is self-recognition in a mirror the same as consciousness of self, as is often assumed, or is this yet another example of us being misled by an ambiguous word—in this case, the very devilish word 'self'? I call it a devilish word because it has a super-seductive fiendish way of insinuating itself into the heart of what we mean by being conscious and tempts us into assuming that anything with 'self' attached to it must be conscious. If we are not careful, the word 'self' bypasses our critical thinking so that self-recognition seamlessly becomes conscious-self-recognition and then morphs into full-blown awareness of self as a person.

However, the ability to distinguish between self and non-self has no necessary connection with consciousness at all. One obvious example is our immune system, one of the major achievements of which is to distinguish 'self' from 'other' so that invading bacteria and viruses are attacked, but the cells of our own body usually are not. This is why organs transplanted from another body are rejected unless special steps are taken to suppress the (normally adaptive) immune response to attack and reject anything foreign. This recognition of self is accomplished by special

[45] Suddendorf and Butler (2013)
[46] Kohda et al. (2019)
[47] Cammaerts and Cammaerts (2015)

cells (T cells and B cells) that respond to proteins used by our own body cells and treat them differently from those coming from other bodies or invading pathogens.[48] So here the distinction between self and non-self is made simply by proteins interacting in certain ways with each other, with no need—or even a hint—of conscious oversight.

Another example is the ability, found widely across the animal kingdom, of being able to distinguish between the effects of self-induced and other-induced movement. When you move around the world, your eyes receive a constantly changing image of the world, but you don't panic that the world is moving and you are in the middle of a large earthquake because as you move, your eyes are effectively warned to expect movement of the visual image of the world due to your own movement. This is called 'reafference'[49] and you can demonstrate it in action by pressing your finger lightly on the side of your eyelid while keeping your head stationary. The external world appears to move. If you were not actively moving, your eye was not expecting movement and so interpreted the change in visual image as a movement in the real world. Normally, this is all done completely unconsciously, so you may not have realized that reafference is a constant feature of your vision. Yet the ability to distinguish self-induced from externally induced movement is a basic property of all movers, allowing animals to recognize real movement in the external world when it happens but not to constantly respond to movement caused by their own locomotion.

Recognizing 'self' from 'not self' is thus not some newly acquired conscious skill but an ancient trait evolved as a consequence of just moving around and repelling invading organisms. It does not, in itself, need consciousness at all.

[48] Dettmer (2021)
[49] Jékely et al. (2021)

What if the answer is 'yes'? (A computer could easily do that)

As we have seen, the point of the question *How easily could a computer do that?* is not to say one way or the other whether animals have conscious experiences like ours. It is simply to get you to stop and think whether a few lines of computer code can produce, without the need to mention consciousness at all, behaviour that could easily be taken as something that could only be done if there was a conscious mind to guide it. The question is intended simply to test the plausibility of claims for animal consciousness. Although many people now use computers, relatively few know how they work or have learnt how to write code. They may therefore simply not realize how easy it is to make a computer do animal-like and indeed human-like things. If you do not know what computers can do, it makes it easy to jump to the conclusion that a complex behaviour can only be done by a conscious mind, simply out of not knowing how easy—and how plausible—the alternative unconscious explanations can be.

Answering 'very' to the question *How easily can a computer do that?* does not, therefore, show that the animal achieves the same thing without being conscious. It simply says that there may be alternative explanations to immediately invoking consciousness. It puts more runners in the race, opens more options, and invites you to consider more possible alternatives.

I hope I have convinced you that whenever you come across a complex behaviour and wonder whether there is a conscious mind at work, it is therefore worth considering two alternative hypotheses. One is that there is indeed a conscious mind behind what you are seeing. The other is that there is a mindless set of algorithms at work. The temptation is to immediately opt for the first hypothesis on the grounds that the behaviour is so complex that a computer could not possibly do that. It also sounds more exciting and more on the side of the animals than so-called kill-joy explanations.[50] But I hope you can now see that the second hypothesis should not be so easily dismissed. Even a small knowledge of coding is enough to show that computers can mimic highly complex behaviour—such as self-monitoring, integration of information, motivational trade-offs, self-recognition—that are often claimed to be the hallmark of consciousness in animals. Not only can they be deliberately programmed to mimic such behaviour but even more interestingly they may spontaneously produce it simply as part of carrying out their own effective functioning. Having to decide between conflicting priorities or deal with information overload, for example, is a common problem with similar solutions whether you are a robot, an animal, or a human.

So, if we really want to understand consciousness and which animals have it, we have to seriously consider the alternative hypothesis that they may be achieving complex behaviour without consciousness. And even if you are already convinced and you are trying to make the case for consciousness in animals to someone else, I think you will see that it would be unwise to argue that you think animals must be conscious simply because they do things that you think a computer could not possibly do. That, as we have seen, is a very weak argument indeed. Computers can do a lot that animals can do and will do even more in the future. The case for consciousness in animals deserves better arguments than an appeal to the (increasingly mistaken) belief in the ineptitude of robots.

The point of this chapter has been to emphasize the importance of considering what can be done without consciousness, using the question of whether the same could be done by an unconscious entity such as an immune system or a computer as a sanity check. Consciousness in animals has not been ruled out, nor has the idea that, one day, we may even find that computers themselves have become conscious. All this chapter has been concerned to do is

[50] Shettleworth (2010)

to inject a measure of caution into discussions about animal consciousness and to provide a brake on too easily accepting evidence that may appear to be more certain than it actually is.

You only have to think of the power that AI already has to fool us with fake images and fake voices to realize the importance of constantly asking questions about how a given result is produced. AI can hold conversations and write books, but it has no understanding of what the words mean. It is just incredibly good at stringing words together in ways that make us think there must be a conscious mind behind what it says and writes. But there is none. Criteria for recognizing consciousness are going to have to be considerably tightened up if we are to keep pace with what computers can already do without consciousness. To make matters even more complex, there will undoubtedly be examples of 'playing the system' or 'gaming consciousness', meaning that for every test proposed for consciousness, someone will deliberately set out to write a computer program that could pass it. Their program might tick all the right boxes as specified in the consciousness 'test' but would not in fact be conscious. Or would it? Would there come a point when we would decide that a machine had passed so many tests and ticked so many boxes that we would have to conclude that it was, after all, conscious? Computers are inevitably going to become better and better at passing such tests, and so making ourselves aware of the different ways in which test criteria can be met without consciousness is going to be as important for our assessment of machine consciousness as it currently is for assessing it in animals.

I cannot emphasize too strongly that the aim of this chapter has not been to disprove consciousness in non-human animals (or machines) but to take us one step nearer being able to tie down what makes consciousness unique and what it seems to be necessary for. By being clearer about what *can* be done without consciousness, we move gradually closer to the critical question of whether there are some things that *cannot* be done without consciousness. We are closing in on what contribution consciousness might make to the working of a brain. We are therefore coming a bit closer to understanding what it is and what it does.

As we have seen, there is a great deal that can be done without consciousness and so a great many abilities that are often associated with it can be ruled out as diagnostic signatures of a conscious mind at work. If we want to identify the genuine tell-tale signs of consciousness, therefore, we have to look beyond things that could equally well be done by unconscious entities such as immune systems or our current computers. We need to ask instead whether there are some tasks or abilities that absolutely require consciousness. In the next chapter, we undertake a search for such consciousness-necessary tasks in the hope that if we knew what these are, we could then move on and ask the next crucial question: Can animals do them too?

6

Unconscious minds

Of all the evidence we might assemble about consciousness, you might think that the most reliable, the most incontestable, and the least likely to be challenged would be our own. Each of us knows we are conscious. We could go further and say that this is perhaps the only thing we can be certain of in this whole uncertain field. Yet it is recent discoveries about human consciousness that, almost paradoxically, provide the biggest challenge of all for the study of non-human consciousness.

Although consciousness seems to dominate our own existence, much human behaviour is in fact carried out quite unconsciously.[1] Many complex tasks that may appear to have the hallmark of consciousness can be done without it, including reading and doing arithmetic,[2] driving a car for short periods, and even learning the rules of a complex game.[3]

Dehaene expresses this idea in quite extreme terms by saying that for most of the time, most of the operations of the human brain are carried out without consciousness at all. *A wild profusion of unconscious processes weaves the textures of who we are and how we act*, is how he puts it.[4]

He argues that, even for us, conscious experiences are rare, and only a few brain operations make it to conscious awareness. What is of particular relevance to the issue of animal consciousness is why this should be the case—why, that is, we humans should be so dominated by unconscious processes. The

[1] Axelrod et al. (2015); Rolls (2020); Cleeremans et al. (2020)
[2] Sklar et al. (2012); Schelonka et al. (2017)
[3] Colman et al. (2010)
[4] Dehaene (2014) p.191

answer goes by the rather off-putting name of 'multiple routes to action', which just means that our brains have many different ways of achieving the same outcome, some conscious and some not.[5]

As the vertebrate brain evolved over millions of years from fish ancestors to land mammals including ourselves, it kept most of its old parts and simply built on top of them as it took on new functions. We literally do carry around a 'fish brain' in our heads, but it is a fish brain to which a great many additions have later been made. In this, the brain is unlike most other parts of the body which have evolved and then been lost if they were no longer needed. The ancestors of horses had five digits on the ends of their limbs, but these were lost over time as their hooves evolved for fast running. The ancestors of blind cave fish had fully functional eyes, but these were lost when darkness made them useless. Brains, however, seem to have kept everything. Brains are like an old house that started life being heated by open wood fires, then used electric fires, then had gas central heating installed followed by heat pumps and solar panels, but instead of throwing out an older system when a newer one was fitted, its successive owners kept them all and, furthermore, still use them all on different occasions. Such an old house would have multiple routes to heating, and brains similarly have multiple routes to behaviour.

For example, we have multiple routes to the single outcome of raising a hand. At the most basic level, we can do this as a reflex. You touch a hot stove and your hand is quickly raised before you are even aware of any burning sensation at all. Only a few moments later, when your hand is well out of harm's way, do you become aware of what has happened and start feeling the pain. Such unconscious reflexes are vital when something has to be done quickly. Consciously thinking that your hand has been damaged and then coming to the conclusion that you should be removing it would

[5] Rolls (2014)

take far too long. Real damage could be done while your conscious brain considered the various alternatives and decided what to do. Much less damage is done by using a swift unconscious reflex action that isn't even reported to the conscious brain until after the danger has already been dealt with.

Reflexes are fast, unthinking and often highly effective ways of protecting ourselves from immediate harm. But they are also clumsy and inflexible. For finer and more coordinated movements, we have several other routes for moving our hands—ones that give us a much wider range of more finely adjusted movement but only some of which involve consciousness.[6] Suppose you were shopping in a supermarket and your conscious thoughts were all about what you were going to buy and how much it was all going to cost. Then you see a real bargain, a very reasonably priced can of beans, say. Thinking consciously only about how you are going to make a bean stew that evening, you reach up, grasp the can, and then put it in your basket. Without consciously thinking about what you are doing, you execute a perfectly controlled movement with just the right extension of your arm to reach the can, the right pressure of your fingers to hold it securely, and a delicately balanced movement to lower the can gently, and without dropping it, into the basket. You did not issue conscious instructions to the muscles of your hand and you probably didn't realize how this was all done, just as people who can very skilfully catch a high-speed ball in mid-air are quite unable to say how they do it.[7]

But while our hands can move unconsciously to do the most amazing things, such as typing at a keyboard or playing a musical instrument, we can also bring our hand movements under conscious control. Trying to thread a needle or learning to play the piano for the first time, for example, requires concentration and consciously thinking about exactly what our hands should

[6] Willingham et al. (2002)
[7] Reed (2010)

be doing.[8] The evolutionary legacy of keeping our various ancestral brains and piling them on top of one another means that we now effectively have different brains for different purposes—some fast and unconscious when immediate action is needed, others slower and conscious when there is time for reflection and rational control.[9]

A particularly dramatic effect of having more than one brain system is provided by the work of Joseph LeDoux,[10] who spent many years working on anxiety-reducing drugs, that is, drugs designed to help people whose lives are ruined by their fear of everyday events such as going out or engaging in social situations. LeDoux and his team initially worked on rats and identified a number of drugs that had a noticeably calming effect on the behaviour of the rats. One of the crucial tests they carried out is called an 'open field test', which involves putting a rat into a large arena and watching what it does. When first introduced to this new environment, most rats will stick to the edges of the arena, defaecate and urinate, and either remain completely motionless or flatten themselves as they run. They try to hide or attempt to escape. LeDoux took all of these behaviours as expressions of 'fear'. (The significance of putting 'fear' in quotation marks will become clear shortly.) He was able to find several drugs that would turn 'fearful' rats into calm rats that ventured out into the middle of the arena, did not defecate, and explored the whole of the open field. His trials were so successful that he received permission to try out the same drugs on human patients. The human trials, however, produced results that surprised even him.

When people took the same fear-reducing drugs that had worked successfully on rats, their fear and anxiety did indeed change for the better, but only in some ways. Their usual signs of fear and worry, such as a racing heart, sweating, and panicky

[8] Montero (2010)
[9] Kahneman (2011)
[10] LeDoux (2014); LeDoux and Pine (2016); LeDoux and Hoffman (2018)

breathing, were all reduced. But in spite of this, they reported *feeling* just as anxious as ever. So, the drugs had successfully dealt with the autonomic signs of their fear but had not been much real help in improving their lives because their conscious experience of fear was as strong as ever. LeDoux[11] berated himself for falling into the word trap of using the one word 'fear' to cover both types of responses. He had assumed that because the drugs affected the outward and visible signs of agitated behaviour in both rats and humans, they would also affect the conscious feelings of being afraid in humans. Instead, he had to conclude that, from what his human patients had told him, the drugs affected the behavioural and physiological signs of fear but not the conscious feelings.

The explanation that LeDoux gave for his results is that there are two different brain routes for fear in humans. One of these routes—the unconscious route—goes through the amygdala, a relatively old part of the brain that is found in both rats and humans. In both species, this route runs via the brainstem and gives rise to a variety of physiological and behavioural symptoms of fear, such as changes in heart rate, breathing, and, in rats, remaining motionless. It was this route through the amygdala that was affected by the fear-reducing drugs.

The second route, however—the one that involved the conscious experience of fear—was not through the amygdala. This brain circuit appeared to go directly to a different part of the brain, the cortex, which was not affected by the drugs. It was therefore still active in the anxious patients despite their physiological symptoms having been reduced. Recent neuroimaging studies on connectivity in the human brain show that this second route goes to the orbito-frontal cortex, a part of the brain known to be involved with the conscious experience of emotions.[12] Furthermore, the human amygdala has much less functional (and anatomical) connectivity

[11] LeDoux (2014)
[12] Taschereau-Dumouchel et al. (2022)

to the neocortex than does the orbito-frontal cortex. The human orbito-frontal cortex, on the other hand, has connections to the parts of the anterior cingulate cortex involved in short-term memory planning, language, and emotional experiences,[13] giving anatomical support to the idea of two largely separate brain circuits.

Rats, by contrast, have a much less developed orbito-frontal cortex than primates.[14] While rats and humans share the older, amygdala-based route to emotional responses, rats have a much less developed second route that, in us, leads us to feel and be able to describe our emotions. This does not of course show that rats are unable to consciously feel fear or other emotions, but it does show that they do not have the same brain systems as we do for doing so. This in turn has obvious implications for how far we can use an animal's behaviour to interpret what it is consciously feeling. Behaviourally, the drugs had comparable effects in the two species. We humans, however, have multiple routes to action and the brain routes we share with rats are not the ones that—in us—are involved with the conscious experience of fear. It follows that we cannot assume that a behaviourally fearful rat is also a consciously fearful rat, even though it might be. 'Fear' has two different routes. Behavioural similarity does not necessarily indicate similarity of conscious experience.

Once, it might have been possible to argue that although we could not know for certain that an animal consciously feels fear, the most plausible conclusion was that an animal that behaves fearfully like us also feels afraid like us. Now, the balance of plausibility has shifted. The existence of separate conscious and unconscious routes to the same action in humans shows that behaviour can and does occur without underlying conscious feelings. This becomes even clearer when we consider the conscious experience of pain.

[13] Rolls et al. (2023)
[14] Passingham (2021)

As mentioned in an earlier chapter, pain is one of the few areas of consciousness studies where enough words exist to make the important distinction between what an animal (or person) *does* and what they *feel*, between anatomy, physiology, and behaviour on the one hand and conscious experience on the other. One word, 'nocioception', refers to the working of a whole set of equipment—sense organs, neural pathways, and brain areas—that we and other animals have for detecting injury or illness, while the other word, 'pain', refers to the unpleasant conscious feeling that something is hurting. Pain may or may not accompany nocioception, but at least by having two different words, we can make it quite clear whether we mean that an animal is just showing a response to injury or is also consciously hurting as a result of that injury.

In mammals including humans, detection of noxious stimuli—nocioception—is usually done by two separate sets of nerve fibres.[15] A-delta fibres are myelinated (wrapped in a fatty sheath) and transmit nerve impulses very rapidly, acting as first responders to tissue damage. The reflex that is triggered by your hand touching a hot stove uses these fast-acting A-delta fibres. C-fibres are smaller in diameter, unmyelinated, and conduct nerve impulses more slowly. Some of them react to both mechanical tissue damage as well as to excessive heat or cold, whereas others respond primarily to either one or the other, giving mammals the ability to discriminate between different sorts of noxious stimuli.[16]

However, once detected, the nocioceptive information is transmitted to many different parts of the brain, which modulate it in various ways and determine how unpleasant a pain actually feels.[17] People with damage to their orbito-frontal cortex, for example, reported the very odd experience that they were aware that there was a painful stimulus but that somehow it did not matter to

[15] Lynn (1994)
[16] Sneddon (2018)
[17] Ossipov et al. (2014)

them.[18] Their brain damage allowed them to keep an awareness of the painful stimulus but removed the unpleasant and intrusive quality that pain has for most people.

Rainville et al.[19] used hypnosis to show how dramatically feelings of pain to the same stimulus can be increased or decreased just by suggestion. They hypnotized willing human subjects and immersed their left hand in a painfully hot (47 °C) waterbath. They then suggested to them, under hypnosis, either that the pain was becoming more and more unpleasant or that it was getting less unpleasant. Despite the fact that the water on their hand remained at exactly the same temperature, the subjects systematically reported that their pain felt correspondingly more or less unpleasant in line with what they were being told was happening. Images of their brain activity (PET scans and EEG) while they were being hypnotized showed that activity in the anterior cingulate cortex (which is closely connected to the orbito-frontal cortex) was mirroring their subjective reports of pain unpleasantness—up when they reported that it was more unpleasant and down when they reported it was less.

The role of the orbito-frontal cortex in the emotional impact of pain was further demonstrated by another brain scan study (this time using fMRI) in which it was shown that both a painful stimulus (a prick with a stylus) and a pleasant touch (velvet) gave rise to more activity in the orbito-frontal cortex than a neutral touch (wood).[20] As we have already seen in connection with the action of anxiety-reducing drugs, rodents have a much less developed orbito-frontal cortex than primates.[21] This means that we share some, but not all, of the multiple routes that nocioceptive stimuli take into the brain and so share some, but not all, of the machinery of pain perception.

[18] Freeman and Watts (1950)
[19] Rainville et al. (1999)
[20] Rolls et al. (2003)
[21] Passingham (2021)

Multiple routes to action—several different brain routes leading to the same behaviour in ourselves—mean that overall similarity between humans and non-humans in behaviour does not on its own provide sufficient evidence for inferring similarity in conscious experience. The human brain is a many-storied edifice, built on and added to over evolutionary time. It retains much of the structure found in our vertebrate ancestors and so inevitably has much in common with the brains of other animals that share our evolutionary heritage. We have not stopped using reflexes just because we have reasoning minds that enable us to decide what to do. We have not stopped using unconscious routes to action just because we also have the ability to be consciously aware of what we are doing. On the contrary, we use everything we have inherited and put it all to use, like the house heated in many different ways. The existence of multiple routes to action in our own brains means that an animal can be 'like us' in what it does, but we are still left with the question of whether it is 'like us' when we are using a conscious route or 'like us' when we are using one of our many unconscious routes. To answer that, we need a way of identifying behaviours that can only be done via a route that we know from independent evidence definitely does involve conscious awareness.

Although there are many things we humans can do without consciousness, it also turns out that there are some tasks that, it has been claimed, we appear to be unable to do unless we are conscious. These tasks are often ones that require us either to combine stimuli in new ways or to retain information over a substantial period of time.[22] By identifying what these 'consciousness-necessary' tasks are for us humans and then seeing if animals can do them too, it has been argued that we at last have a more solid way of attributing consciousness to animals.[23] If they can do things that we can only

[22] Dehaene and Naccache (2001)

[23] Ginsburg and Jablonka (2019); Birch et al. (2020); Droege et al. (2021); Dung (2022); Mason and Lavery (2022)

do with consciousness, then does it not follow that they must be doing them consciously too? Let us see whether this approach can at last give us the insights we need into animal consciousness.

One task that has been proposed as being consciousness-necessary is called trace conditioning. Trace conditioning just means that there is a delay between the time when a previously neutral cue (such as a tone) occurs and when some unpleasant consequence (such as a puff of air to the eyeball) happens. The first stimulus, therefore, has to be remembered for long enough for there still to be a memory 'trace' of it when the second stimulus comes along. Otherwise would be impossible to make the connection between the two.

Now, having a puff of air squirted onto your eyeball is not particularly pleasant. And people (and animals with eyelids) tend to blink quite naturally if this is done to them. If a tone reliably indicates that the air puff is about to happen, then people learn to start blinking when the tone is heard, even without the air puff. So far, nothing remarkable. Just standard classical conditioning as described by Pavlov with his salivating dogs, apart from the fact that the tone has to leave a longer memory trace than if the two events happened simultaneously. This is clearly not a problem for most animals as trace conditioning is found right across the animal kingdom, including in sea slugs[24] and fruit flies.[25] It is also extremely easy to give a computer a stream of data and get it to pick out correlations between events with varying time delays. No need to suggest consciousness at all.

However, by introducing a slight twist to the way trace experiments are conducted, it has been claimed that trace conditioning then becomes a consciousness-necessary task. Clark and Squire worked with a group of human volunteers and played them a tone followed by a puff of air to their eyeballs (either 500 ms or

[24] Glanzman (1995)
[25] Dylla et al. (2013)

1,000 s later).[26] The twist was that at the same time as this was happening, everybody was shown a silent movie, intended to distract them from noticing the connection between the tone and the air puff. The result was that many people were so totally distracted by watching the movie that they did not notice any connection between the tone and the air puff at all. They failed to show any evidence of trace conditioning and did not learn to blink when they heard the tone.

After the experiment, the participants were debriefed about what they thought had been going on. The ones who had failed to make the connection between the tone and the subsequent air puff reported that they had not consciously realized that there was any connection between the two. On the other hand, the participants who had learnt to blink when they heard the tone and in anticipation of the air puff also reported that they had consciously realized that the tone was associated with the unpleasant air puff. From this, it was concluded that consciousness was necessary for the trace conditioning to happen. Since there was no trace conditioning in people who did not report being conscious of the connection between the tone and the air puff and everyone who reported being conscious of the connection showed successful trace conditioning, it seemed that being conscious was indeed necessary for successfully learning the connection between the tone and the delayed air puff.

This suggestion was taken up as a possible way of establishing consciousness in non-human animals by Droege et al.[27] Specifically, they argued that it is not so much trace conditioning by itself that indicates consciousness but whether or not the trace conditioning can be prevented by switching attention away from the trace-conditioning task using some sort of distraction. If an

[26] Clark and Squire (1998)
[27] Droege et al. (2021)

animal shows trace conditioning under one set of conditions but not when it is distracted by other incoming stimuli, then, they argued, this suggests that the animal is showing that, like humans, it is conscious when it learns successfully.

However, this is to confuse 'attention' with 'consciousness'. It is well established that we and other animals are limited in how many stimuli we can process at any one time and that we deal with this problem by switching between different information streams at different times.[28] Our sense organs are constantly receiving information about vision, hearing, taste, smell, and touch in far greater quantities than our brains can possibly process and we ignore much of it and only process the most interesting or the most urgent. Here is yet another example of how words can change the way we think. If we call this process 'selective attention', then that immediately implies that we are switching our consciousness from one focus to another, since consciousness and attention seem so intimately bound up with one another. However, if we call it 'selective filtering', we can see that what we are talking about is simply focussing our limited brain resources onto specific pieces of information at any one time. This may, of course, involve consciously experiencing this switch, but it could just be one way that any system with limited processing capacity—animate or inanimate, conscious or not—deals with information overload. It is therefore not surprising that if someone is looking at a video, they may not be processing information about tones and air puffs as efficiently as if they were attending exclusively to how one predicted the other.

Indeed, the Clark and Squire result confirms just how limited our own brain's processing capacity really is. Some people failed to learn even the simple trace-conditioning task if they were watching a film. But the reason this happens is not because we are conscious beings but because we have brains with limited processing capacity

[28] Zentall (2005)

and we have to filter out some information and process only part of what is coming in. In the case of human beings, we know that this ability is sometimes accompanied by a conscious experience of 'attending' (hence the common conflation of attention with consciousness), but there is now increasing evidence that even for us, our attention mechanisms can also operate unconsciously.[29] For example, if a stimulus is flashed so briefly that you do not see it, you will still become more 'attentive' and faster at responding to other stimuli at the same location.[30] You may not be aware that your attention has been switched to the 'interesting' location, but at some unconscious level, it has become more likely that you will in fact respond to other things in the same location.

It is therefore essential to separate 'attention' (focussing the brain's limited resources), which can be done either consciously or unconsciously from 'consciousness' (being aware of what is being focussed on).[31] Selective attention sorts relevant from irrelevant incoming information but is itself no guarantee that there is a conscious mind to pass the relevant information on to. A completely unconscious processor with similar informational limits would have the same difficulty in dealing with two tasks at once. Given that animal brains are just as or even more limited than our own, we should expect that they too would find it more difficult to cope with learning a trace-conditioning task if they were being distracted by something else at the same time. If they fail in the trace-conditioning task under these circumstances, it does not necessarily mean that they are conscious when they successfully show trace conditioning. It just means that they can do it better when their brains (whether having conscious experiences or not) have not been overloaded with too much information.

Trace conditioning as a way into animal consciousness is thus not as clear-cut as it might seem. Trace conditioning itself is not

[29] Dehaene (2014)
[30] McCormick (1997)
[31] Lamme (2003)

difficult for a computer to do and does not carry any necessary implications of consciousness. The fact that successful trace conditioning only occurred in the Clark and Squire experiment when participants could report that they were consciously aware of the connection between two stimuli could of course mean that consciousness was necessary for such conditioning to occur. But this is not the only possible explanation. It could also just show that a brain with limited processing capacity has to focus on the learning task for this to be successfully accomplished, leaving it open as to whether that focussing (attention) occurred with a conscious stream or without it. Trace conditioning, even when it can be eliminated by a distraction, is therefore not a signature hallmark of consciousness, human or animal. It is just what is to be expected of any brain or processing unit of limited capacity that is forced to ignore some of the barrage of incoming information it is receiving.

The second example of apparently conscious-necessary abilities we will discuss comes from experiments that exert very careful control over the *duration* of the stimulus that people are exposed to. You may remember from Chapter 4 that psychologists have consistently found that a visual stimulus has to be presented for at least 60 ms for it to be reliably seen consciously.[32] If a stimulus is presented for less time than that, it may have some effect on behaviour, but people are likely to say they have seen nothing. It is therefore possible to present the same stimulus but to render it either supraliminal (consciously seen and reported on) or subliminal (unconscious with a denial of having seen anything) to a subject depending on how long the subject is allowed to see it.

This technique has now been used to try and identify tasks that humans can only perform when they are conscious of having seen the stimuli in question—that is, when the stimuli are presented for more than about 60 ms. One such task is a kind of learning

[32] Dehaene and Naccache (2001)

called reversal learning. This involves a participant (which could be either human or animal) not only learning a task but also what to do when circumstances change, and specifically when the reward contingencies are reversed. A participant first has to learn a simple discrimination, such as to press a black rather than a white panel to obtain a reward. Then when this task has been thoroughly mastered, the experimenter reverses the contingencies so that now pressing the white panel gives the reward, whereas pressing the black one does not. When that task has been learnt, the contingencies are reversed again, with pressing the black panel now being rewarded and so on. As this pattern of constantly reversing continues over time, some participants may begin to realize that there is a pattern to the rewards: whenever a previously rewarded panel is suddenly unrewarded, pressing the opposite panel gives an immediate reward. The really clever participant can theoretically get so good at this that they just need one trial with no reward to reverse their responses. They learn a reversal 'rule': after one non-reward, change. In contrast, a participant that has not learnt this reversal rule will keep having to relearn from scratch after each reversal and show little or no improvement over time.

Reversal learning is indeed a rather special sort of learning because to do it successfully, it demands the ability not just to learn to do one thing rather than another but to be able to see a longer-term pattern in what is going on and to change behaviour appropriately. As a technique, it provides clear objective evidence that a participant has detected a long-term pattern even when this is not said in so many words. If, after extensive training, an animal learns to switch responses immediately after it fails to get a reward only once, then it has clearly learnt the reversal rule. If it shows no progressive reduction in the number of mistakes it makes after each reversal, then it has clearly not learnt the rule.

Now, on the face of it, although reversal learning is more complex than a straightforward learning task, it is not something that demands a very great intellect. Machine learning finds it a relatively simple task, for example.[33] However, it has been linked to consciousness because of empirical evidence that it is one of those tasks where humans appear to need to consciously see the cues in front of them to be able to do it successfully.

For example, Travers et al.[34] studied human subjects in a reversal learning experiment under the two conditions—conscious and unconscious—brought about by the method of varying the duration of stimulus presentation. A subject sat in front of a computer screen and at the beginning of each trial, little arrows (« or ») would appear on the screen. The participants' task was to predict on which side of the screen a larger target (an X) would appear a few moments later, using the direction of the arrows to guide them. They then had to press a left or right button. Sometimes the arrows would point towards the correct side so that the subjects could follow the direction of the arrows to predict the appearance of the target, but then once they had learnt this, everything would be reversed and the arrows would point in the opposite direction from the correct target position, so the correct response became the side that the arrows were pointing away from. Once they had learnt this change, everything would be reversed again for another block of trials and so on. Each subject did the experiment twice—once with the little arrows flashed for just 33 ms (unconscious condition) and once with the arrows visible for 400 ms (conscious condition). The results showed that only when the subjects were consciously aware of the direction of the arrows did they show evidence of reversal learning. In the condition in which they saw the arrows for 400 ms and reported that they saw them, they could learn to reverse the connection between arrow direction and target position and flip their responses quickly from arrows pointing towards the target position to away from it and back again as the experiment progressed. But if the stimuli were presented for such a short time

[33] Ganin et al. (2016)
[34] Travers et al. (2018)

that the subjects reported they had seen no arrows at all, they were unable to do this. They had to relearn after each reversal as if they had never encountered the change before.

Travers et al. concluded from their results that in humans, consciousness is necessary for this example of reversal learning and furthermore that consciousness actually facilitated the learning of the reversal rule. A number of authors took this to show that if an animal shows evidence of reversal learning, then it, too, is conscious.[35] Since many animals, including rats, pigeons,[36] fish,[37] and frogs,[38] show reversal learning in the sense of learning to change their behaviour when contingencies change, they are therefore, so the argument goes, showing evidence of consciousness. At first sight, this sounds highly plausible. By anchoring this conclusion firmly to carefully controlled experiments on humans where verbal reports are possible, it would seem that we at last have a way of navigating safely to identifying consciousness in animals where verbal reports are not possible. Or do we?

A major problem arises from the use of stimulus duration to move humans in and out of conscious awareness. In the human experiment described above, the difference between the conscious and the unconscious parts of the experiment was achieved by manipulating the length of time for which the subjects saw the little arrows, 400 or 33 ms. However, a longer stimulus is also a larger stimulus and might also provide a better basis for unconscious processing as it provides more information.[39] Just because in the human experiment the longer stimulus resulted in the perception of the arrows reaching consciousness, it does not follow that unconscious processing then stops. Unconscious processing might also benefit from the longer stimulus, even to the point of

[35] Birch et al. (2020); Mason and Lavery (2022)
[36] Rayburn-Reeves et al. (2013)
[37] Parker et al. (2012); Kuroda et al. (2017); Fuss and Witte (2019)
[38] Liu et al. (2016)
[39] Green et al. (1957); Swets et al. (1961)

now being able to learn the one-trial reversal rule. In other words, even for subliminal stimuli, there can still be an improvement in reversal learning tasks if stimuli are presented for longer simply because there is more information in a longer-lasting stimulus than in one that only lasts for a shorter time.[40] In the animal experiments quoted above as evidence for reversal learning, the stimuli were all available for very long periods of time—seconds or even minutes—unlike the few milliseconds of the Travers study on humans. In typical experiments on fish, for example, the animals swim into an experimental chamber and so have sight of the choice stimuli for a relatively long time before they have to make their choice. The point is that unconscious processing could also have benefitted from this longer exposure.

Unconscious processing of visual information has indeed been shown to improve the longer a stimulus lasts. In one recent experiment, people were briefly shown photographs of faces and asked to say whether they thought the person was happy, sad, or neutral.[41] The photographs were presented for extremely short times—8.3, 16.7, and 25 ms—that is, all within the subliminal duration zone that is supposedly too short for the participants to consciously see anything. The probability that someone could correctly identify the emotional expression on a face was only 12.2% with the 8.3 ms presentation but rose to 36.8% with the 16.7 ms presentation, still below the usually reported threshold for conscious perception. Interestingly with the 25 ms presentation of faces, the correct responses rose to 85% and many of the participants reported that they had actually seen faces or thought they might have done, suggesting that there is not a sudden cut-off from unconscious to conscious vision but a gradual improvement with stimulus duration, even when an image is not consciously seen. The longer the stimulus, the more information it provides.

[40] Rolls (2004)
[41] Schräder et al. (2023)

Now we can see why the method of manipulating human conscious perception by these ingenious masking experiments should not so easily be used to infer consciousness in animals. Experiments in which humans are shown stimuli for varying lengths of time do indeed show that some tasks can only be performed if a stimulus lasts long enough to enter consciousness. However, contrary to what is often claimed, such experiments do *not* show that if a stimulus lasts long enough to enter consciousness, then unconscious processing no longer happens. Indeed, we can expect that unconscious processing too might become more effective than it is with a very brief stimulus. Longer, louder, brighter (more salient) stimuli are easier to discriminate and easier to detect from background noise than shorter, quieter, dimmer stimuli whether the receiver is a human, an animal, or a radar device.[42] The greater power of longer stimuli is to be expected quite independently of whether something enters conscious awareness or not.

A series of experiments by Ben-Haim et al.[43] attempted to address this problem directly. They devised some ingenious tests that could be done both by humans and rhesus macaque monkeys. One of their experiments was a variation on the reversal learning tests described above and used the same technique of flashing images on a screen and carefully controlling stimulus duration so that humans either reported seeing them or not seeing them. They devised experiments that were as similar as possible for humans and monkeys as subjects. Both species were fitted with eye-tracking devices that showed where they were looking. For both, the rewarded response was to fixate on the correct stimulus for at least 200 ms. In one of their experiments, the participants had to view a screen on which two different images appeared, one on the left and one on the right. Just before the targets appeared, a small * would be flashed onto the screen, either for 250 ms

[42] Swets et al. (1961)
[43] Ben-Haim et al. (2021)

(human conscious condition) or for just 17/33 ms, which, as we have seen previously, is usually not consciously seen by humans. The participants had to learn to fixate their eyes on the side where the correct target was about to appear. To make the task a bit more difficult, the * indicated not which of the two subsequent target images was going to be on the rewarded side but which was going to be on the unrewarded side—the one that should not be looked at. Apart from not being able to ask the monkeys what they saw, humans and monkeys did essentially the same experiment. The result, as we might expect from previous results, was that both species were more likely to choose the correct target when the * had been presented for 250 ms and when the humans reported that they had consciously seen it than with the 17/33 ms when most of the humans said they had not.

Aware of the objection that these results might be due to the greater salience or information content of the longer cue stimulus, Ben Haim et al. attempted to control for this by doing a further experiment in which they tested a new set of human subjects with the same subliminal (17/33 ms) presentations of the * before showing them the two target stimuli. As expected, the subjects did not perform very well and did not report seeing anything on the screen before the targets appeared. Then, half-way through the experiment, the subjects were told about the presence of the quick-flashing * and told that this might help them to solve the problem. Many (67%) of them then said they could now see the *, learnt the 'choose opposite' rule, and started scoring almost as highly as subjects in the superliminal (250 ms) condition. Ben Haim et al. argue that this 'suggests that cue awareness rather than cue saliency (which remained constantly difficult to see, here at 33 ms) is the critical factor in performance on our studies'.

Of course, although the cue 'remained constantly difficult to see', its saliency to the individuals who were now told of its significance did not. The subjects now knew that the uncertain image of something that they might previously have dismissed as background noise was now the key to solving the task. This information clearly had the effect of lowering their threshold of response to the * enough for it to enter consciousness, but this lowered threshold could also have made their unconscious processing more effective too.

To summarize, experiments that carefully control how long a stimulus is presented have shown that there are some tasks that humans are unable to do unless they are shown stimuli for longer than about 40–60 ms and, moreover, report that they can see them. These are the experimental facts, which are repeatable and robust. It is the interpretation of the facts that we need to question. There are major implications for understanding both human and non-human consciousness. Firstly, it is important to note that these findings do *not* rule out unconscious processing at exposures longer than 60 ms. They show that a stimulus usually has to be seen for at least 40–60 ms for someone to report that they have seen it and also that there are some tasks that can only be successfully carried out when stimuli are presented for long enough to reach conscious awareness. This has been widely interpreted as showing that conscious awareness is necessary for humans to perform certain tasks.

However, it does not follow that just because a stimulus presented for more than 60 ms is consciously seen, unconscious processing suddenly stops and lets the conscious processing take over. On the contrary, unconscious processing, like conscious processing, is likely to become more efficient as stimulus duration is extended and the salience of the situation becomes greater. Where the consciousness/unconsciousness boundary is manipulated by altering stimulus duration, it is therefore not correct to conclude that consciousness is necessary for a particular task. It is quite possible that the same task could be carried out unconsciously, provided the unconscious system was also given a bit more time to view the stimulus.

Secondly, this has major implications for how we interpret animal experiments. Just because the monkeys in the Ben Haim experiment showed a difference in learning ability depending on whether stimuli were presented for a very short duration (17/33 ms) or a longer duration (250 ms) does not demonstrate that the monkeys were experiencing a similar shift from an unconscious to conscious state because purely unconscious processing can also be expected to improve with longer stimulus duration. Longer stimuli are easier to process both consciously and unconsciously. This is even more important to keep in mind when considering other experiments on animal reversal learning that have been used as evidence of animal consciousness. These experiments invariably involve very long exposure times, often of several seconds or even minutes—much longer than the human threshold for consciousness. Humans need an exposure to stimuli of at least 40–60 milliseconds to show reversal learning and to report consciously on what they are doing in a very specialized kind of experiment. However, this says nothing at all about the consciousness or otherwise of animals able to view stimuli for several seconds or even minutes. All of us—animals and humans—might be able to do reversal learning unconsciously if given enough time. The controlled stimulus duration experiments with their emphasis on deliberately presenting stimuli for very short periods of time do not allow us to rule out this possibility.

I will leave the reader to make up their own mind about using the results of these various human experiments to infer consciousness in animals. My own view is that they do not capture the unique working of consciousness—even in humans—nearly as convincingly as is often claimed. Manipulating humans in and out of consciousness by varying stimulus duration certainly tells us about what can and cannot be achieved with shorter and longer stimuli. It tells us how long a stimulus needs to last in order to be consciously seen and reported. But it says nothing about the way in which unconscious processing might also be varying with stimulus duration, in parallel with the conscious stream. Reversal learning, like trace conditioning, has not yet been shown to be a 'consciousness-necessary' task for humans and so provides much weaker evidence for consciousness in animals than has been claimed.[44] This is, of course, disappointing but perhaps what we can learn from this is that a fruitful way forward for the study of animal consciousness would be to look for other tasks that are genuinely conscious-necessary in humans and where enhanced unconscious processing can be more definitively ruled out. If animals could also do those same tasks, then we might have more convincing evidence for conscious experiences in them too. As it is, we have no such tasks obviously available, and so how best to study consciousness without verbal reports remains a big, as yet unsolved, issue.

I want now to turn to a third inference that is sometimes drawn from the experiments we have been discussing, an inference that calls into question what they really tell us about human consciousness itself. Travers et al.,[45] in their study of reversal learning, concluded:

> *Taken together, these results suggest that although some information can be processed and affect behaviour without reaching conscious awareness, effective rapid learning, at least on this particular task, within the time frames we investigated of up to 100 trials requires that information to be represented consciously, that is, consciousness facilitated learning.*

Saying that consciousness facilitated learning might seem only trivially different from saying that consciousness is correlated or associated with improved reversal learning, but there is, in fact, a world of difference. To say that consciousness is correlated with

[44] E.g. Birch et al. (2020)
[45] Travers et al. (2018)

learning leaves open what the causal pathway might be, but to say that it facilitates learning implies that consciousness is somehow involved in the neural pathways leading to better learning. Now, although it may not be immediately obvious, that is a claim with momentous implications for what consciousness is.

Think about it for a moment. Can consciousness really be playing a *causal* role in the neural machinery of the brain? Can consciousness actually cause neurons to fire? Or does it intervene in synapses and disrupt action potentials? If we allow consciousness this role, does this not begin to violate the basic laws of physics and chemistry?

Sooner or later, everyone who takes consciousness seriously has to face up to this awkward question of the exact relationship between consciousness and the brain—specifically how brains give rise to consciousness and whether consciousness affects how brains work. Sooner or later, everyone has to decide for themselves where they stand on this issue that philosophers call the 'mind–body problem' but is just as well called the consciousness–brain problem. Whatever we call it, it is one of the big questions still to be answered about consciousness, both animal and human. It is sometimes thought of as separate from the question of who is conscious, which, as we have seen, can be thought of as part of the 'problem of other minds'. In practice the mind–body problem and the problem of other minds are intimately related. We want to know who is conscious (which are the other minds), so we first need to know what sort of brains give rise to conscious minds and how they do it (mind–body problem). As we look around the animal kingdom for evidence of consciousness, we need to know what we should be looking for. What is the difference between a brain with consciousness and one without it? How does any sort of brain give rise to consciousness? If we understood that, we would be in a much better position to know who is conscious.

As we will see in the next chapter, however, the mind–body problem has been around literally for centuries and there are still no satisfactory answers. It was raised in the form we currently know it by a French mathematician and polymath nearly four hundred years ago and continues to be the subject of one very long argument that, even now, it is impossible to ignore.

7

It's hard not to be a Dualist

René Descartes (1596–1650) was an extraordinary man. He unified algebra and geometry and introduced the system of Cartesian (*x*, *y*) coordinates we still use today. He put forward theories of perception, of the origin of the universe, of the refraction of light, and of the nature of matter. He thought deeply about how he knew that anything really existed and came to the famous conclusion 'cogito ergo sum' (I think therefore I am). He also started an argument about the relationship between consciousness and the brain that has been raging ever since.[1]

The reason why the views of a philosopher who lived 400 years ago are still relevant now is not because he was right (most people think he was quite spectacularly wrong) but because he split 'mind' and 'body' apart with such force that they have never since been satisfactorily put back together again.[2] Almost everything that has been said or written on the subject of mind and brain since Descartes can be seen either as an attempt to show that he was wrong or to find a way in which just conceivably, with many qualifications, he might have been right. Even now, in the twenty-first century, scientists and philosophers who want to put forward their own theories about consciousness often feel compelled to make it clear that they definitely are not (or definitely are) Cartesians. It has become almost a mark of intellectual rigour to spell out a position either for or (more usually) against Descartes. Open almost any modern book about minds and brains and there you will find

[1] Descartes (1637, 1649)
[2] Frith and Rees (2007)

him—sometimes staring out from a portrait with large, doleful eyes, sometimes present just through a description of his challenging ideas. Descartes has been part of the way people understand and misunderstand the relationship between mind and body for nearly four centuries.

It is a remarkable achievement for any scientist to have their ideas still actively talked about after such a long time, not as history but as part of current scientific debate. It is even more remarkable when those ideas are demonstrably incompatible with the known facts of anatomy and physiology, as Descartes' are. Descartes' legacy was not to be correct, but to be wrong in a very logical and rigorous way. By setting out what he saw as a complete explanation of the relationship between mind and body, he threw down a challenge. The challenge was not to show that he was wrong. That was easy, even with the knowledge of anatomy available at the time. The challenge was, and still is, to find a convincing alternative explanation about how the mind and the brain interact. Descartes described exactly how he thought this happened, even down to giving precise anatomical details about where in the brain he thought the interaction took place. Few people since have done this in such detail. Few have come up with such a complete physiological explanation of what happens when mind meets brain.

So Descartes' challenge remains, almost taunting us down the centuries. If we are so sure he was wrong, then what is our supposedly 'better' explanation? In this chapter, we will see why it has proved so difficult even for his severest critics to completely rid themselves of his powerful and seductive ideas and therefore how careful we must be in specifying how brains give rise to consciousness. Descartes' views—ostensibly despised and discarded—still shape the way people think today about the minds of both animals and humans. Being able to simultaneously answer both Descartes and his critics is one of the best ways of deciding what your own views about consciousness are.

Princess Elisabeth's question

The most controversial view that Descartes left us with is that the universe is made up of two kinds of substance—mind and matter. Because he insisted there are two kinds, this idea is called Dualism, and is usually referred to as Cartesian Dualism in acknowledgement of his contribution to the idea.[3] Descartes argued that most of the universe—including the planets, the sun, and the bodies of all living things—is made up of material substance or 'matter', which exists in three-dimensional space and operates deterministically according to the universal laws of physics. But he also proposed that there is a second kind of substance, 'mind' or soul, the essence of which is thought. Unlike matter, mind is immaterial and cannot be located in space or measured in terms of its length, weight, or temperature. Human beings, Descartes believed, are a union of both matter and mind. Our material bodies are made of matter just like the rest of the universe so that processes such as the digestion of food and the beating of a heart can all be understood in mechanical terms as the working of a machine following the laws of physics. But we also have immaterial minds that allow us to think rationally and then to influence and actually interfere with the working of the body machine by making our rational thoughts guide our actions. We therefore rise above being just machines through our capacity for thought and the power of our will to influence what we do.

Descartes' distinction between a machine-like body and a separate mind concerned with thoughts and feelings has immense appeal as a description of human existence. It perfectly describes our intuitive view of ourselves as a private subjective 'self' living inside a publicly observable body. It satisfies our desire to be more than just machines and to believe that we have free will and even a soul. It seems such a natural distinction that it slips easily, often unnoticed, into almost every discussion about consciousness. Even

[3] Descartes (1637)/1996

talking about 'the mind–body' problem presupposes a two-way split, a duet, an acknowledgement that we are talking about two different things whose connection has yet to be explained.

The reason we remember Descartes today, however, is that he did not stop at just formulating an intuitively plausible idea. He was not content with a vague, hand-waving account that the mind has 'something' to do with the body. He went on to develop a novel theory of how mind and brain interact. He proposed that it took place in the pineal gland, a small structure situated in the base of the brain. He believed that this tiny gland had the capacity to act as a two-way intermediary between mind and body. He was goaded into developing this extraordinary idea by the persistent criticisms of his many correspondents, and most particularly by one of them, Princess Elisabeth of Bohemia.

Princess Elisabeth was the eldest daughter of Frederick V, a deposed King of Bohemia and a granddaughter of James I of England. On her own initiative, she started writing to Descartes in 1643 and they corresponded regularly until his death in 1650.[4] Elisabeth told Descartes that there was a serious flaw in the argument that mind and body were separate substances if, at the same time, he was also trying to argue that the mind could influence the body. If there were two kinds of 'stuff' in the universe—matter stuff and mind stuff—and one is operating according to the laws of physics and the other is not, how, she asked, could one possibly influence the other? The idea of a rational but immaterial mind causing a change in a material body made intuitive sense, but it raised a whole new set of problems that, as Elisabeth pointed out, Descartes had simply failed to explain.

Elisabeth's criticism spurred Descartes into providing a detailed mechanism for how such a mutual interaction could happen. He fixed on the pineal gland as the place where the interaction occurred because, unlike many other anatomical structures that

[4] Shapiro (2013)

occur in pairs, the pineal gland is a single organ, ideally suited, therefore, as a single unified conduit. It is also situated in the brain and so is well placed both to receive incoming sensory messages and to pass messages from the brain to the muscles. In his book *Passions of the Soul*,[5] which incidentally he dedicated to Elisabeth, Descartes proposed an elaborate theory of how the pineal gland achieved this remarkable feat. When the sense organs are stimulated, he argued, vibrations travel up nerve fibres into the brain and by an elaborate system of valves in the walls of the brain ventricles, an image of the sensory stimulus then appears on the surface of the pineal gland. It is this image that Descartes believed then makes the conscious perception of the stimulus available to the rational mind. On the way back, once a rational action has been decided on by the mind, the pineal gland vibrates, this time in response to the emotions and passions that the mind has generated as a result of its thinking. These vibrations of the pineal gland in turn lead to the opening of pores in the brain ventricles (a network of fluid-filled cavities) that allow animal spirits to flow down hollow nerve tubes and so activate the muscles. The mind thus does not change the way the body reacts directly—it cannot force the body to fight, for example—but it can change the pineal vibrations from those that cause the body to run away to those that cause it to stay and fight. The mind can only indirectly, but still definitely, affect the material body.

Descartes' explanation is, of course, completely refuted by modern science. Muscles do not work by animal spirits coming down hollow tubes. Vibrations of the pineal gland are not key to the emotions. But the paradox is that, despite its absurdity, it is the level of detail with which Descartes described his proposed mechanism that is what makes it relevant today. If he had simply stopped at saying that there were two kinds of substance, mind and matter, but had not specified exactly how they could be connected, then he

[5] Descartes (1649)

would be one of a long line of philosophers, biologists, and others who have speculated freely but then failed to provide a mechanism or to answer the interaction question that Princess Elisabeth so perceptively raised. Even though Descartes' own answer was wrong, he gave us a benchmark against which all subsequent attempts to explain how mind and brain are connected have come to be judged. Any complete theory of the relationship between mind and body has now first to specify whether there are two kinds of substance, or just one or many. If there is more than one, then the theory has also to specify the exact mechanism by which the two are supposed to interact. If there is only one, then an explanation has to be provided for why the mind manages to give such a vivid impression of being separate.

So we can now see why it is that Descartes is still relevant to discussions of 'mind' and 'body'. His ideas about a vibrating pineal gland can be easily dismissed as ridiculous, but his clear grasp of the mind–body problem and what a complete mechanistic account of consciousness would have to explain remains as important as ever. Descartes was a true scientist in that he made predictions that could be tested publicly and proved wrong. The fact that he was wrong in no way detracts from the importance of his attempts to specify exactly how mind and matter might interact. And his answer to Princess Elisabeth's question, flawed though it was, has still not been adequately replaced by anyone else.

Objections to Dualism

The philosopher Gilbert Ryle famously ridiculed Descartes' belief in 'mind stuff' as equivalent to believing that there is 'a ghost in the machine'.[6] The neuroscientist Antonio Damasio expressed what he thought of Descartes' views by giving his 1994 book the title

[6] Ryle (1949)

of *Descartes' Error*.[7] Most philosophers and scientists nowadays would strenuously deny that they believe in Dualism as Descartes expressed it (sometimes also called Substance Dualism, because mind and body are seen as two separate substances). Immaterial mind stuff is too much like magic for modern science to accept it as a serious proposition about the way the world is constructed. What is fascinating, however, is how difficult it has been, even for Descartes' severest critics, to throw out Dualism altogether. Many people who claim to reject it outright turn out, when really interrogated, to still cling to the idea that 'mind' is somehow different in substance from 'body'. Overt Dualists are relatively rare, although some eminent scientists such as John Eccles have 'come out' as such.[8] But inadvertent or camouflaged Dualists, who simultaneously ridicule Dualist ideas while unwittingly falling back on them, are much more common.[9]

For example, many people have speculated about the evolutionary functions of consciousness, that is, they have asked what advantage a conscious organism would have over one that was not conscious.[10] An easy answer is to argue that having a conscious experience of, say, pain helps an animal to escape from something that is injuring it and the memory of that pain then stops it from repeating a dangerous action again in the future. The evolutionary 'function' of conscious pain is thus as a help in escaping from and avoiding injury.

But this easy answer leads straight into the Dualist trap.[11] Natural selection has certainly favoured early responses to danger and the withdrawal of limbs at the first sign of injury, but that can be very efficiently done by an unconscious nocioceptive system. If an organism has fully functional mechanisms for avoiding injury that

[7] Damasio (1994)
[8] Eccles (1992)
[9] Dennett (1991)
[10] Griffin (1976); Merker (2005)
[11] Polger (2006)

can be described in objective physiological terms without mentioning conscious experience, then what extra use is there for a pain that consciously *hurts*?

Here is the trap and the reason why it is so necessary for you to think carefully about what you believe if you want to avoid being a Dualist. Natural selection cannot favour an animal that consciously experiences pain over one that does not unless that conscious experience makes a difference to the effectiveness of its ability to respond to injury or its ability to learn to avoid dangerous situations. The animal that experiences pain must in some way be better at surviving and reproducing than one that just has an effective nocioceptive system combined with an ability to learn. To use our previous test, an unconscious robot can easily learn to avoid certain locations or actions if they have damaging consequences. A driverless car, for instance, can learn from experience how to avoid crashing into things. But if, in talking about animals, the conscious experience of pain-that-hurts is claimed to be necessary over and above the workings of the objectively observable nocioceptive and learning mechanisms, then the door has been opened to a very Dualist interpretation of what is happening.[12] If something extra, beyond objectively observable neurophysiological events, is being used as part of the explanation of the mechanism of how animals and people avoid being injured, then something very like Descartes' 'mind stuff' is being surreptitiously added to the explanation. To ask what is the function of pain, with the implication that it has a separate function from that of neurophysiological mechanisms that are correlated with it, is to risk being a very Descartes-like Dualist. It seems so natural to think this way, and some effort may be needed to decide whether that is really what you believe.

People who cannot feel pain provide a case to think about. Such people are at a great disadvantage because they cannot feel when

[12] Polger (2006)

they have injured themselves, so a minor graze becomes worse and they end up being severely damaged.[13] They are, of course, living proof of the importance of a fully functional nocioceptive system. A Dualist might see them as people with nervous systems exactly like everyone else's, just with an extra bit (i.e. the conscious pain bit) missing. A strict non-Dualist, on the other hand, will point out that their nervous systems must be different in important respects from people who can feel pain. Many such people do in fact have genetic mutations that affect the transmission of nerve impulses from the body surface to the brain so that their brains do not receive the message that there is an injury.[14] They therefore have a failure of nocioception that is correlated with their inability to feel pain. The damage they suffer shows very clearly how important it is to have a properly functioning nervous system that can convey injury messages to the brain. But it says nothing at all about any additional functional significance of the pain that accompanies injury in most people.

If you think that the conscious experience of pain has a function *over and above* that of neurophysiological mechanisms alone, then you may be a Dualist without realizing it. You would effectively be saying that pain is caused by 'something extra' that intervenes in the normal workings of neurons and brains and that, without it, there would be no conscious sensation of pain. That 'something extra' sounds suspiciously close to Descartes' mind stuff, with all the problems that has caused down the centuries.

Now we can see one reason it is so hard not to be a Dualist. Dualism itself has an intuitive appeal that is hard to shake off. It appears to fill a gap in our understanding of what consciousness is and to explain what we have no other explanations for. The other reason is that, as we shall see in the next section, the alternatives to it seem so unappealing, so inadequate and sometimes just plain wrong.

[13] Mogil et al. (2000)
[14] Goldberg et al. (2007); Drissi et al. (2020); Eagles et al. (2022)

The disappearance and reappearance of 'mind'

The obvious way of avoiding Dualism is to say that there is only one kind of 'stuff' and that the matter in the physical world is all there is. Mind or consciousness would then become nothing more than the workings of the physical, material brain so that once neuroscience has told us how the brain works, it will also have told us all we need to know about how the mind works. There will be nothing 'extra' to discover, no magic ingredient that neuroscience has failed to describe because what we call mind or consciousness is purely a product of the only substance that exists—physical matter. This view is accordingly called Monism or more commonly, Physicalism or Materialism.[15]

The case for Physicalism is that it has no place for magic mind stuff, and so it appears to be much more scientific than Dualism with its notions of ghosts, spirits, and souls. Physicalism sees the brain as a physical object operating within the normal laws of physics and chemistry and so makes the study of consciousness look scientifically respectable. But the problem is that Physicalism doesn't really explain mind at all. By insisting that the working of the brain is all we need to know about and indeed all we can know about, Physicalism avoids the mind–body problem altogether by effectively insisting that there is no problem. In a Physicalist world, there is no room for any extra private, subjective experiences that cannot be studied by objective means. Mind cannot be separated from body.

Yet this flies in the face of our everyday experience of what it is to be alive and to be human. We know, from living inside our own bodies, that there is more to being in pain than just the activation of nerves that could be recorded externally. We know, at first hand, that pain *hurts* and so the private, subjective sensation of being in pain cannot be just dismissed as a non-problem. Pain matters.

[15] Stoljar (2016)

We can accept that the sensation of pain is linked to a known chain of physiological processes that occurs when we injure ourselves and that there are special nerve fibres and parts of the brain that are associated with our body's response to injury. But the private subjective experience of feeling pain for ourselves is very different from the outer world in which other people's behaviour and physiology can be observed and studied objectively. Your pain is like my pain because similar objectively observable nervous pathways are activated. But your pain is also different from my pain because only you can experience your hurt and only I can experience mine. However much we know objectively about each other's bodies, something is always apparently missing. We are missing each other's private conscious experience of pain.

We therefore seem to be left with an apparently impossible choice. Either we see mind and body as two separate substances, leaving ourselves open to accusations of believing in non-scientific mind stuff and having to agree with Descartes that our minds do not follow the same laws of physics that govern everything else in the universe. Or we say there is no such thing as mind stuff, keep strictly to what is objectively knowable by scientific methods and then find ourselves unable to 'explain' the one thing that each of us thought we knew with absolute certainty from our own experience—that we have minds and are consciously aware of the world around us.

A number of different ways of escaping this Dualism dilemma have so far been tried.

Escape by denying mind altogether

The first way of avoiding Dualism is outright denial. Everything to do with the mind, such as feelings or thoughts, is simply deemed to be unscientific and therefore not worthy of further consideration. Science deals with the observable and the testable, and consciousness is neither observable nor testable. It is therefore outside science, and, from a scientific point of view, it does not exist. This was the way consciousness was dealt with by the Behaviorists that we discussed in previous chapters. By denying consciousness, John Watson saw Behaviorism as a revolutionary new way of making psychology into a true science, in contrast to its previous reliance on introspection and untestable ideas from psychoanalysis. Consciousness 'is merely another word for the "soul" of more ancient times', he wrote. He added, 'The old psychology is thus dominated by a subtle kind of religious philosophy'. His contempt for this 'old' psychology is obvious.[16]

Behaviorism offers an escape from Descartes' Dualism by denying the existence of 'mind stuff' altogether. At the same time, it answers Princess Elisabeth's awkward question about interactions between mind and body. If there is no mind, there are no mind–body interactions to worry about and the focus can be entirely on the workings of the physical brain. Descartes' outdated ideas about animal spirits and vibrating pineal glands can be replaced with the much greater understanding we now have of how nerves, muscles, and the brain itself actually work. Even more importantly, the normal scientific standards of replicability and hypothesis-testing can be applied in much the same way as they are in other branches of science, making the whole enterprise fully scientific.

The problem is that something vital is now missing. In giving such priority to logic and objectivity, Behaviorism turned its back on, but did not solve, the problems of consciousness. Calling consciousness unscientific does not make people doubt the reality of their own conscious experiences. It does not make them suddenly give up wondering what consciousness is, or which other animals have it, or wanting an explanation of how brains give rise to it. All it does is to convince them that science is not up to the job of providing the answers they want.

[16] Watson (1929)

So eventually dissatisfaction with Behaviorism set in as its limitations became more and more obvious and the subsequent revolt against it also triggered a complete rethink of the interaction between mind and body. If the dilemma about how the two are related could not be solved by the Behaviorists' solution of denying the existence of 'mind' altogether, then some other way of avoiding Dualism was going to be needed. Consciousness was still a problem and how it related to the physical brain was still just as much of an issue as it had been for Descartes. The proposed alternative was even more radical.

Escape by claiming that mind and brain are the same thing

The post-Behaviorist solution that initially gained the most support was called Identity Theory.[17] Unlike Behaviorism, Identity Theory recognizes the existence of conscious experiences but then says that these conscious experiences are identical to the physical processes happening in the brain. States of consciousness are not just correlated with states of the brain. Mind states *are* brain states. Conscious states of being in pain *are the same as* the brain states underlying them. There is nothing else to pain except what the brain is doing. Mind is still there, but it is identified with the body it is in.

Identity theory provides another answer to Dualism but gives the disconcerting impression of being just a clever conjuring trick. At the beginning of the trick are two entities—mind and body—and a serious question of how the two can connect with one another. Then a wand is waved, and suddenly there is only one, and all the worries about causal connections have simply vanished. But, as with watching a skilled conjuror, we are left with the feeling that things are not quite as they seem. Surely, it is not possible to solve

[17] Smart (2000)

the long-standing problems about mind and body just by defining them out of existence. Defining mind and body as the same might look like a solution, but it didn't feel right. For many non-philosophers, it was even difficult to understand what it meant to say that mind states were the same as brain states when they seemed so obviously different, one private and subjective, the other public and objective.

Other cracks began to appear in Identity Theory too. If conscious states were identical to human brain states, then it would seem to follow that only humans could be conscious, not birds, fish, or octopuses with very different brains. It seemed unsatisfactory to rule out, by definition, the possibility that non-human animals and even machines could be conscious. It should surely be an open question where consciousness exists, not something to be ruled out in advance because, by definition, it could only happen with a human brain. So a revised sort of mind-into-body incorporation was proposed, which emphasized not so much what brains are made of but what those brains are doing.

This revised version was called Functionalism.[18] Mind and body are still seen as the same (to avoid Dualism), but instead of identifying mind with anatomical states of the brain, functionalists identify mind with the functions that a brain carries out. Just as the wings of birds, bats, insects, and aeroplanes are very different but can all carry out the function of 'flight', so functionalists see consciousness as more a property of what a brain (or a machine) can do than what it is made of.

One obvious consequence of Functionalism is that it opens up the possibility that conscious experiences are a property of a much wider range of brains than just human ones. Even having no nerve cells at all and being a man-made machine would not rule out consciousness, provided that the right processing was being carried out.

[18] Levin (2004)

And that, of course, is both a strength and a weakness. While Functionalism potentially welcomes many non-human entities into the consciousness club by not making possession of a human brain a condition of entry, its failure to specify the conditions of membership leaves the door wide open to all sorts of entities, including insects, plants, and machines. Differences of opinion over specifying what functions distinguish conscious entities from unconscious ones (such as we have discussed in previous chapters) have led to it now being possible to claim that anything at all is conscious, including plants and all living things. Such a free-for-all, 'anything goes' attitude to consciousness is both unsatisfactory and unscientific.

Escape by 'something in parallel'

The third alternative to Dualism is the view that mind and body are both real; each has its own identity, but they run in parallel. Each happens in its own way and at the same time as the other but without a causal link between them. So a thought happens at the same time as a brain process, but it is somehow separate from it—a sort of add-on or extra.

This slightly disconcerting idea of minds and brains running in parallel but not causally connected has its origins in the nineteenth century. Thomas Henry Huxley ('Darwin's bulldog') proposed a theory he called Epiphenomenalism, derived from the Greek 'epi' meaning 'above' and the late Latin/Greek 'phainomenon' meaning 'to appear'. Mind appears because of brain activity, but the mind is somehow above what the brain is and has no way of influencing it. Huxley used the analogy with a steam train and its whistle to explain this idea. 'Consciousness ... would appear to be related to the mechanism of the body', he wrote, '... ... simply as a (side) product of its working, and to be completely without any power of modifying that working as the (sound of) a steam whistle which accompanies the work of a locomotive is without influence on machinery'.[19]

For Huxley, the causal connection was one way only. The steam engine made it possible to blow a whistle with the excess steam that would otherwise be wasted, but the blowing of the whistle had no effect on the engine. Similarly, a brain gives rise to conscious thoughts as a by-product of performing certain operations, but those thoughts have no effect on the brain.

Eighty years later, another word for 'above', this time using the Latin 'super', was used to coin a new term that also implied that minds and brains could run in parallel without necessarily being causally connected. The new word was 'supervenience'.[20] Supervenience means that as a conscious mind moves from one thought to another, there are accompanying changes in the brain that happen at the same time, that is, they come above or in addition to the brain activity. Conscious thoughts 'supervene' on what the brain is doing and are inextricably linked in ways that have yet to be explained. Supervenience, like Epiphenomenalism, is thus a compromise idea that conscious events are related to brain states and that they are indeed very closely tied to them. However, it avoids the Dualism problem of implying that a non-physical mind can actually interfere with the physical workings of the brain.

Some versions of supervenience agree with Huxley that brains causally influence the working of the mind but not the other way round, whereas other versions of supervenience go further and deny any causal connection in either direction. But whatever version of supervenience is adopted, the brain is seen as the prime mover in that there can be no change in states of consciousness without corresponding changes in the state of a physical brain. This grounds consciousness firmly in the physical world, but it still does

[19] Huxley (1874)
[20] Davidson (1970)

not explain what consciousness is or why some brain states are accompanied by conscious experiences and others are not. It could even be said to replace one mystery with an even greater one. What can it possibly mean to say that two things always occur together but are not causally linked? Why should they always occur together if nothing binds them?

Escape to different levels of explanation

Not being a Dualist is hard because it means having to answer the question 'what causes a brain to give rise to conscious experiences?' without invoking Dualistic mind stuff. That in turn seems to mean that the only 'causes' that are available to answer this question are the basic processes that neuroscience has uncovered—the propagation of nerve impulses, transmission across synapses, increased blood flow to active areas of the brain, and so on. So, to avoid being Dualists, we would thus appear to have to avoid explanations that invoke consciousness as a cause of such brain events because this would allow a non-physical extra to interfere with what is known about the physical functioning of the brain. Metaphorically speaking, such explanations would allow consciousness to stick its non-physical fingers into purely physical chains of events and cause changes in processes such as the flow of ions across nerve cell membranes or the release of transmitters between cells. And that would be back to Dualism.

Avoiding Dualism, therefore, depends on not allowing consciousness to be used as a cause of the way the cells in the brain operate and yet, as we have seen, describing brain events purely in terms of physiological events appears to leave out what we are trying to explain—consciousness itself. A recent suggestion for a way out of this dilemma is to see consciousness and brain activity not as having any causal links in either direction but as the same phenomenon described at two levels of explanation. Understanding this distinction depends critically on what is meant by something being a 'cause' of something else.

The most widespread definition is that A is said to be the cause of B if the occurrence of A is a necessary prior condition for the occurrence of B—that is, B does not happen unless A has happened first. Causation is established by a combination of statistical correlation and experimental intervention.[21] What this means in practice is that A and B must be observed to be correlated or reliably occur together in time and that, crucially, this correlation must remain intact in the face of experimental attempts to disrupt it. So, we might notice a correlation between a cock crowing and the arrival of daylight, but the cock does not cause the sun to rise, as can be shown by preventing the cock from crowing and showing that, nevertheless, daylight still comes anyway. Well-designed experiments can be thought of as systematic attempts to disrupt correlations, usually by setting up conditions in which A is prevented from occurring to see whether B still occurs. If without A there is no B and the correlation no longer exists, that would suggest that A is indeed a necessary cause of B.

However, even if the correlation between A and B still survives experimental intervention, this would not distinguish between 'cause' and 'supervenience' that figures so prominently in philosophical discussions of the relationship between mind and body. On at least some definitions, a supervenient relationship would also survive experimental intervention and yet would not, by definition, be causal. This means that something more is going to have to be added to the definition of 'cause' if we are to be clear what it means. Edmund Rolls has recently suggested that what we should add is precedence in time—that is, something cannot be a cause of something else unless it happens before it.[22] To say that A causes B would thus imply that A occurs and then, some time

[21] Woodward (2005)
[22] Rolls (2021)

later, B occurs.[23] Most people would agree that something cannot be a cause of something else if it occurs after it, but nor can it, Rolls argues, if it occurs simultaneously. The time interval between two events can be very short, but unless there is some measurable delay, A cannot be said to be a precondition or cause for B. If two things happen at exactly the same time, neither can be the cause of the other, and it is more likely that they are actually the same thing but expressed at different levels of explanation. Rolls argues that this is the most accurate, non-causal way of describing the relationship between consciousness and the brain.

For example, if you were consciously thinking about dolphins, you might have a train of thought that causally linked one idea to another, perhaps starting with thoughts about the sea, conservation, or fishing nets that then in turn trigger or cause thoughts about the welfare of dolphins. This is the higher level of explanation. At the same time as you are having these thoughts, however, your brain will be very active, with neurons firing and setting off activity in other neurons to which they are connected. This is a lower level of explanation. The important point is that these two chains of activity will be going on at exactly the same time, with no time delay between them, so on Rolls' view, one cannot be the cause of the other. Individual neurons at the lower level do not wait for a top-down signal from the higher thought level to tell them what to do next.[24] Causality operates within a level, not between levels so that the networks of neurons that make up the brain have their own causally complete interactions that give rise to network-level properties that can, at a higher level, be described in terms of recognition of a face, a memory, or a decision.

This level-of-explanation approach potentially provides one way out of the centuries-long debates about how mind and body might be causally related. It does not, of course, solve the hard problem

[23] Granger (1969)
[24] Rolls 2021.

of how consciousness arises in the brain, but it does clear some clutter out of the way, clutter in the form of seemingly intractable arguments about causal links between mind and brain.[25] It at least allows us to talk about conscious experiences without invoking non-scientific ideas about mind stuff or trying to believe that consciousness operates outside the laws of physics or having to conclude that consciousness somehow mysteriously intrudes into the behaviour of nerve cells.

You may not agree with this way of looking at the problem of consciousness, but if you are ever tempted to use consciousness as a causal explanation for something a brain is doing—such as improving an ability to learn or feel pain—then it is worth asking yourself whether you really mean to imply that immaterial consciousness is causing events to happen in a very material brain. Do you really believe that non-physical consciousness is affecting the way molecules and brain cells work? If that is what you believe, then of course you are perfectly entitled to do so, but you should realize just how radical—and how Dualist—you are being. You would be implying that consciousness somehow alters the interactions between neurons or changes how different brain regions interact. You would then have to explain exactly what consciousness is doing. Is it stopping neurotransmitter molecules from crossing synapses? Is it reducing or amplifying action potentials so that nerve cells behave differently?

The alternative is to see consciousness not so much as a rogue agent let loose inside the brain causing nerve cells to misfire but something more like what happens when a computer is running a video game. While the game is in progress, the computer moves from one line of code to the next (software-level change), and this is simultaneously realized at the hardware level by a series of changes in the state of the computer's logic gates. There are no missing links at either level that can only be described by appeal

[25] Kim (2011); Polger (2006)

to the other level. Software changes only cause software changes and hardware changes only cause hardware changes. They occur together because they are simply different levels of explanation for the same events. The actions of a fictional player character in a video game do not cause changes in the computer's logic gates because causal links only occur within, not between, levels of explanation.

On this view, the next step in the search for who is conscious just became a little clearer. We need to focus on the right level of explanation.

The brain is not a computer—but it computes

There is an obvious flaw in describing levels of explanation in the brain as being like the software and hardware levels in a computer, which is that the brain is not a computer. Well, it is certainly not like the digital or analogue computers we have now. It is much slower and much less accurate; it has all sorts of constraints on how it can function because of the way neurons connect to one another. It is organized differently, and, of course, it is made out of very different materials.[26] But there is also a sense in which the brain very definitely is a computer. It computes. It performs calculations, it processes information, it stores memories, it extracts patterns. For example, when the image of a person's face is received by the eyes initially, this is just a pattern of light and colour falling on a screen (the retina). The retina does not recognize faces or anything else, but as the information is passed up the optic nerve and into the brain, information about what is being looked at is gradually extracted by a series of operations or computations. In the primary visual cortex (VI), there are neurons that respond to very simple stimuli, such as lines or edges, so that if these respond,

[26] Brette (2022)

this indicates some of the properties that the image must have but not exactly who or even what someone might be looking at. Further up the visual pathway, however, in the inferior-temporal cortex, there are face recognition cells—neurons that are specialized to respond to faces and form part of an ensemble that recognizes particular people.[27] Through several successive operations, the brain turns information about simple stimuli (such as lines and edges) into complex information about whose face it is and, furthermore, can recognize that face regardless of its apparent size, position, and lighting conditions. This is a major computational feat, taking place as a series of successive operations, each one delivering more information about the identity of a face than the one before.[28]

It is, therefore, quite reasonable to call the brain a kind of computer and to talk about different levels of explanation of how it works. It has a software level of calculating, memory, and other mental events that is equivalent to the software code on a computer, and it has lower levels of brain activity equivalent to the changing states of computer hardware. Now, if we wanted to know how a computer performs its operations, we would not learn very much by opening it up and seeing what was going on inside. Even if we had a way of knowing which parts were active at any one time—say, if little lights flashed when different parts were active (the equivalent of brain imaging)—we would not be much the wiser. A sequence of flashing lights would tell us very little about what operations the machine was carrying out. For that, we would want to know how it was programmed, that is, what software code it was running.

Similarly, when trying to understand how a brain works, it may be highly instructive to look not just for the neuronal hardware that underlies consciousness (the Neural Correlates of Consciousness or NCCs) but also for the equivalent of the software code the brain

[27] Perrett et al. (1982); Rolls (2012)
[28] Rolls (2021)

is running when someone has conscious experiences (the Computational Correlates of Consciousness or CCCs). Brain imaging can show that a particular part of the brain, such as the hippocampus, is a hot zone for memory, but it does not tell us how the brain's memory actually works. To do that, we need to go into much more detail about what a brain does when it stores a memory and what it then does when a memory is later recalled. In other words, we would need to know the *computations* that the brain is performing.

To understand the relationship between consciousness and brain activity, therefore, it would seem logical to say that in the future, the best insights may well come from searching not just for the neural (hardware) signatures of consciousness but also for the computational (software) signatures of consciousness. Some computational routes will accomplish great end points without consciousness, like those in current chess-playing computers and Large Language Models, but others, like the ones humans use to play chess or describe their experiences, will involve flashes into conscious awareness. If we could identify what these conscious computations are, we would be a step closer to understanding consciousness itself. At the very least, it suggests that it might be worth looking for what computations (CCCs) the human brain is carrying out when people report that they are having conscious experiences. This would add enormously to our understanding of what consciousness is. It would make sense of our growing understanding of which parts of the brain are involved in consciousness and the kinds of activity that are associated with conscious experiences. It would enable us to see *why* certain parts of the brain are involved in consciousness and *what* operations are being performed.

The hardware and the software of animal minds

Looking for CCCs would also open up new ways of approaching animal consciousness. If we understood the high-level, software, CCC in ourselves, we might be better able to look for something similar in other species, even those that have nervous systems and brains that are quite different from our own. Just as the same software calculations can be performed on different hardware (abacus, difference engine, analogue computer, iPhone), so might the same consciousness code run on totally different nervous systems—molluscan or insect ones, for example. The key question then becomes not *Could* a dolphin/crab/tortoise/oak tree do this or that? Rather, it becomes *How* do they do it? What computations are running in their brains? Are they the same computations that occur when we are conscious, or do they achieve the same results in some other way, without the need for the consciousness code? Are they like chess-playing computers, beating us with their superior algorithms, without the need for consciousness, or are they, like us, computing consciousness itself?

The same questions might even allow us to make real progress with the problem of whether machines can be conscious too. If we understood the brain computations that underlie human consciousness, there would be no reason why these same computations could not be replicated on machines that would then plausibly have conscious experiences too. If you argue that this will never be possible because there is something unique about brain cells that enable them, and them alone, to support consciousness, then be warned, you may be an inadvertent Dualist. You would have to explain what is so special about brain cells that give them this unique property, and I suggest you would find it very difficult to explain what this special property is without invoking something very like Descartes' mind stuff. If you argue that conscious brains evolved by natural selection and this makes them different from any machine, then you would have to explain why artificial selection on man-made machines would be incapable of achieving the same result. If you argue that no machine could ever be conscious because it would always lack that non-physical something that living organisms have, then you would, indeed, be a Dualist.

It is much more likely, however, that future studies of how the human brain works will make such arguments even more difficult to maintain than they are now. As we begin to uncover the CCC in humans, there will be little to stop similar algorithms being implemented on machines. Once this has happened, it will then be difficult to argue that those machines are not also consciously feeling sad or anxious or seeing the world in rich colours. They may not have evolved slowly by natural selection or used networks of nerve cells to achieve their conscious state, but they will surely challenge the very basis of our understanding of who is conscious.

8

Where are we now?

It was signed by over 250 philosophers, neuroscientists, ethologists, animal welfare scientists, and others from around the world. It directly addressed the question of which animals have the capacity for conscious experience. It gave a clear answer. The New York Declaration on Animal Consciousness was issued on 29 April 2024. It stated that there is 'at least a realistic possibility of conscious experience in all vertebrates (including reptiles, amphibians, and fishes) and many invertebrates (including, at minimum, cephalopod mollusks, decapod crustaceans, and insects)'.[1]

This declaration was at least slightly more cautious than the Cambridge Declaration of 2012 mentioned at the beginning of the book, which, you may remember, stated it as a scientific fact that birds, mammals, and octopuses have the 'neurological substrates' for conscious experiences. Neither declaration, however, provided evidence. In both cases, we were simply invited to be impressed by the number of signatories and their scientific or philosophical credentials.

On 16 June 2024, the BBC posted an article by Pallab Ghosh,[2] their science correspondent, entitled *Are animals conscious? How research is changing minds*. The mind-changing research referred to in the title included many of the findings we have discussed in this book such as animals being able to recognize themselves in mirrors, to remember past events, to seek out pain relief, and to prioritize one course of action over another. As with the New York Declaration, a major point of the article was to emphasize

[1] New York University (2024)
[2] Ghosh (2004)

that it is not just birds and mammals that can do these things but fish, crabs, and insects too. Lars Chittka, who has done pioneering work on the behaviour of bees, was quoted as saying 'Given all the evidence that is on the table, it is quite likely that bees are conscious.'[3]

Taking the evidence on animal consciousness even further was a paper by Mark Budolfson, Bob Fischer, and Noah Scovronick, published a year earlier in the prestigious journal Science.[4] They started by making a list of the various kinds of evidence for animal consciousness that we have been discussing in this book, including whether an animal's behaviour is modified by pain killers, whether it shows reward-based learning and trace conditioning, whether it shows flexible self-protective behaviour, parental behaviour, fear-like behaviour, and so on. Their list consisted of forty-seven different kinds of evidence for consciousness in animals, and they then argued that the list could be used to give an indication of an animal's ability to experience suffering (its 'welfare potential'), based on a weighted aggregate score of how many of the forty-seven criteria it fulfilled. So, on this reckoning, a pig comes out with a higher welfare potential score than a python because it fulfils more of the consciousness criteria, but both pigs and snakes come out with a lower score than a human. By giving all species a score on this single scale, the authors claim that this gives an objective way of judging the relative goodness or badness of different policies that affect humans and other animals, making it much easier to make rational decisions that affect more than one species. What this paper shows is that research findings that might have originally been described as 'possible' or 'plausible' evidence of animal consciousness are now being taken seriously as the criteria for the capacity to suffer, so much so that they have actually been combined mathematically

[3] See also Chittka (2022)
[4] Budolphson et al. (2023)

into a single measure of how much one animal might suffer relative to another and proposed as a basis for political and legislative decisions.

I cite these three examples because they make it clear just how much attention is now being paid to the question of which animals are conscious and also, perhaps more importantly, how the scientific evidence that is available is actually being used. To be fair, neither the New York Declaration nor the BBC article says that animals such as insects definitely are conscious. They talk about it as a 'realistic possibility' rather than as a conclusion. The stated aim is more to open up discussion and to consider what is most plausible in light of new evidence on animal behaviour. That can only be a welcome development, but it serves to highlight the need to be ever more critical of the evidence that there is.

There is certainly no lack of interest in animal consciousness. The barriers to thinking and asking questions about it fell away back in the last century and have long since disappeared even from scientific circles. The problem is no longer that no one is studying animal consciousness or is afraid to suggest that some or even most non-human species might have conscious experiences. The problem is now the opposite. It has become one of a too-ready acceptance of tentative evidence that may not stand up to scrutiny. Ideas about which animals are conscious that are initially proposed as 'plausible' all too easily become accepted as definite fact. Unless, that is, it can be shown that there are other equally plausible explanations around.

The aim of this book has been to provide those alternative explanations, not, as I have repeatedly said, to claim that non-humans are incapable of having conscious experiences, but to search for the best possible evidence that some of them might indeed have very vivid ones. If a plausible argument that, say, insects are sentient goes unchallenged, then people come to believe that this is the only explanation for insect behaviour around, and they will start to accept it as fact. But if it is made clear that there are other explanations that they might not have thought of, then the whole picture changes. The plausible suggestion is seen for what it is—a plausible suggestion, not an uncontested fact. The important point is that the evidence for insect sentience, if it survives the scrutiny, might even turn out to be stronger than ever.

The real answer to the 'Where are we now?' question, however, is that currently we still do not know. We cannot give a convincing evidence-based answer to the question of who is conscious because the concept of consciousness itself is still so elusive and beyond what we currently understand.

Despite what it may seem, however, this is not such a bad conclusion to have reached. The questioning that has gone on in the previous chapters has, I hope, achieved something very important. I hope it has shown you that the evidence about who is conscious has turned out to be much more fragile than it might have seemed at first. This matters because being able to put a measure of certainty or uncertainty on what you believe matters. It particularly matters when there are serious social, economic, or personal consequences of turning those beliefs into laws that everybody must obey.

Politicians, like businessmen and most other people, want certainty. They want to 'follow the science', but they don't want to follow the uncertainty that a truly scientific approach brings with it. They want to be told exactly what is going to happen so that they can formulate a vaccine policy, a climate policy, or a business strategy and claim the decision is based on scientific evidence. In the case of animal consciousness, regulators want to know, one way or the other, whether a given kind of animal is conscious so that it can be confidently included or excluded from an animal protection law with the cast-iron guarantee that the decision is based on scientific evidence. But this demand for certainty where none exists has potentially disastrous consequences if the doubts and problems are not fully taken into account.

There are parallels here with medical research. If a controversial drug were to be licensed for public use despite conflicting evidence about its side effects, we would be horrified if the licensing authority were to pretend it was safer than it really was. We would expect the side effects to not only be documented but also positively looked for and any uncertainties associated with the drug's use to be fully explained to patients before it is administered to them. We would expect that there would be follow-up studies about what happens to people who take it after it is released for public use. In other words, in the case of new medicines, we accept that the licensing authorities may have to make important decisions in the absence of all the evidence they might ideally like to have, but we expect that in doing so, they tell us how certain or uncertain they are. We expect to know that the flaws, the objections, and the unsuspected side effects will be exposed because this is what maintains public confidence in medical treatments. It is when the problems with drugs are not looked for or, worse, when they are found and then suppressed by vested interests that public confidence begins to falter. Uncertainty, doubt, and raising all possible criticisms are thus not the drivers of destruction or ways of getting valuable medical treatments banned. They are the route to better treatments, in fact, the only way of getting things right.

So it is with consciousness in animals. Questioning the evidence, even when it seems very plausible, is the only sure way of eventually being able to decide who is conscious. Maintaining an awareness of the difficulties of studying consciousness is particularly important where people adopt the Animal Sentience Precautionary Principle, which, as you may remember from Chapter 2, was stated by Jonathan Birch as:

> *Where there are threats of serious negative animal welfare outcomes, lack of full scientific certainty as to the sentience of the animal in question shall not be used as a reason for postponing cost-effective measures to prevent those outcomes.*[5]

[5] Birch (2017)

The third paragraph of the New York Declaration is revealing in this respect as it clearly invokes the Precautionary Principle: 'when there is a realistic possibility of conscious experience in an animal, it is irresponsible to ignore that possibility in decisions affecting that animal. We should consider welfare risks and use the evidence to inform our responses to these risks'.

Note, however, that in taking this line, the Declaration is subtly using an ethical argument (the precautionary principle) to bolster its scientific one. It is irresponsible, we are told, to ignore the risks to animal welfare by not taking account of the new scientific evidence on animal consciousness. Now, who would want to be held up as being on the wrong side of animal welfare? We are effectively being told that it is better to be on the side of virtue and accept the evidence that animals are conscious than to risk the alternative of being thought not to care enough about animal welfare. This is not a very subtle attempt to persuade us to take a particular view of animal consciousness by implying that we would be irresponsible to think otherwise.

However, as discussed in Chapter 2, while a precautionary, benefit-of-the-doubt argument might be appropriate for making ethical decisions about how to treat different animals, it has no place in scientific discovery, where doubt is deliberately used (or should be) to question what is commonly believed. In other words, there may be enough evidence for ethical or legal decisions, but that is a quite separate question of whether we truly understand animal consciousness from a scientific point of view. You may agree with the aspirations of the Declaration to open up the discussion about which animals are conscious (you may even have signed it yourself!), but that should not put an end to scientific discussions about the nature of consciousness and who has it.

There are many, many things we still don't understand about consciousness, and our attempts to understand what we don't know—which has to include raising doubts about any evidence that may come to light—should not be hampered just because some people have decided that there is enough evidence to use it to make ethical or legal decisions. Declarations, like creeds, are valuable statements of belief, but they are not science. Indeed, where their intention is to imply that people who disagree with them or question the evidence are ethically in the wrong or irresponsible, they may be positively anti-science.

I hope that this book has gone some way to providing a bit of clarity and some guidance as to how to respond in the face of the many claims that are currently being made about where consciousness may be found. It has raised doubts about a lot of the evidence that has been cited but only with the intention of clearing the ground for future investigations that may, one day, enable us to give a more rigorous answer to the question of who is conscious than we have now. I will try and summarize what I hope you will take away from the book:

1. The problem we want to solve is that of phenomenal consciousness—the basic awareness that sometimes accompanies what we and other animals do. This is not easy to define and there may be different kinds of consciousness. The awareness of hunger, pain, or the sensation of light all feel different from each other and different again from more intellectual states such as having a bright idea about how to solve a puzzle. We may have to be content with a variety of different explanations for these different states, rather than one overarching explanation for consciousness.
2. A persistent problem in the study of consciousness is that we often use the same words to describe both the outward signs of states such as 'fear', 'emotion', or 'thinking' and also the inner conscious states that may or may not accompany them. This often makes it unclear whether we are genuinely talking about conscious experiences or just behavioural and physiological changes. One contribution we can all make, therefore, is to be clear (even to the point of being thought

pedantic) whether we are talking about changes in body states, changes in conscious states, or both.

3. The study of consciousness in humans has progressed beyond recognition from being a neglected or even forbidden subject for much of the twentieth century to a recognized science with objective data based on brain imaging, precise control of stimuli, and, above all, the recognition that what people say they are experiencing can be used as scientific data. Verbal reports play a key role in the study of human consciousness.
4. The study of consciousness in species that cannot tell what they are experiencing in words, however, is more complex than it might seem. On the one hand, we have their 'body language' of behaviour, sounds, and gestures to tell us what they are experiencing. Actions often speak louder than words, even with other people. But on the other hand, establishing that there is a conscious mind behind these actions is not straightforward. In particular, we humans are known to have both conscious and unconscious routes to the same behaviour, and much of our own brain activity does not reach consciousness at all. As much of our own behaviour can take place unconsciously, it is difficult to rule out the same for other species. They may be 'like us' in much of what they do, but are they like us when we are acting consciously or acting unconsciously?
5. Most, if not all, of the criteria that have so far been suggested for consciousness can be met, or will plausibly soon be met, by machines, using relatively simple computer code. These include integrating information, self monitoring, and being able to handle motivational trade-offs and so on. This does not rule out consciousness in animals, but awareness of what computers can do (and often do very easily) provides a sanity check and helps to avoid over-hasty conclusions that the criteria that have been put

forward for recognizing consciousness can only be met by conscious beings.

6. A recent development in the study of human consciousness has been to attempt to find tasks that humans can only perform successfully if they are consciously aware of what they are doing and then to ask whether animals can do them too. If they can, this argument goes, it suggests that the animals must also be consciously aware. However, this argument often rests on misinterpretations of the human experiments or on invalid comparisons between the human and animal experiments. In particular, human experiments that manipulate consciousness by varying stimulus duration do not rule out the possibility that unconscious processing also becomes more efficient with longer-lasting stimuli. Contrary to some of the conclusions that have been drawn from such experiments, there are as yet no 'consciousness-necessary' tasks that animals can be shown to do too.
7. It is very easy to slip into a Dualistic way of thinking and start implying that consciousness is somehow altering brain states or vice versa. This can be avoided by seeing brain changes and trains of conscious thought as the same events requiring different levels of explanation.
8. Future advances in the study of consciousness are likely to come from understanding the computational 'signatures' of consciousness, or Computational Correlates of Consciousness (CCCs). This can initially be done in humans, where we have the advantage of being able to ask people what they are experiencing, but may eventually provide a way of looking at the way animal brains operate too. Understanding what operations a conscious brain performs may help us find consciousness, if it exists, in non-humans since the software of the mind may run on many different kinds of hardware (nervous systems).

9. The Precautionary Principle—to err on the side of assuming that a given animal is conscious even in the absence of definitive evidence—may have a role in ethical and legal decisions, but can hinder scientific progress.
10. The fact that there are already in existence laws and regulations that make the assumption that animals are conscious is no reason to stop raising questions about the evidence on which such an assumption is based. On the contrary, the more people assume that evidence is more certain than it really is, the greater the importance of being critical and putting forward alternative hypotheses. Better evidence on animal sentience can only benefit animal welfare.

These ten conclusions can be further reduced to the following essential messages to take away:

- Define words carefully and don't 'flirt' with consciousness unless you mean it
- As a sanity check, ask: How easily could a computer do this?
- Respect the power of the unconscious—it does more than you might think
- Try not to be a Dualist, even though it's hard work trying not to be
- Never let ethical decisions (or Declarations) stop you from asking questions
- Doubt and criticism are essential steps to a clearer vision of consciousness

I hope that you have enjoyed this exploration into animal minds despite its emphasis on what we still don't know and will see it as a positive contribution to thinking more clearly about what remains, for me and for many others, the most intriguing question in the whole of biology. We still have to ask: *Who is conscious?*

Bibliography

Adamo SA. Do insects feel pain? A question at the intersection of animal behaviour, philosophy and robotics. Anim Behav. 2016;118:75–79. https://doi.org/10.1016/j.anbehav.2016.05.005

Adamo SA. Is it pain if it does not hurt? On the unlikelihood of insect pain. Can Entomol. 2019;151(6):685–95. https://doi.org/10.4039/tce.2019.49

Anderson JR, Gallup GG. Mirror self-recognition: a review and critique of attempts to promote and engineer self-recognition in primates. Primates. 2015;56(4):317–26. https://doi.org/10.1007/s10329-015-0488-9

Axelrod V, Bar M, Rees G. Exploring the unconscious using faces. Trends Cogn Sci. 2015;19(1):35–45. https://doi.org/10.1016/j.tics.2014.11.003

Baars BJ. A cognitive theory of consciousness. Cambridge University Press; 1988.

Baars BJ. The conscious access hypothesis: origins and recent evidence. Trends Cogn Sci. 2002;6(1):47–52. https://doi.org/10.1016/S1364-6613(00)01819-2

Balleine BW, Dickinson A. Goal-directed instrumental action: contingency and incentive learning and their cortical substrates. Neuropharmacology. 1998;37(4-5):407–19. https://doi.org/10.1016/S0028-3908(98)00033-1

Barron AB, Klein C. What insects can tell us about the origins of consciousness. P Natl Acad Sci USA. 2016;113(18):4900–08. https://doi.org/10.1073/pnas.1520084113

Barttfeld P, Uhrig L, Sitt JD, Sigman M, Jarraya B, Dehaene S. Signature of consciousness in the dynamics of resting-state brain activity. P Natl Acad Sci USA. 2015;112(37):E5219–E20. https://doi.org/10.1073/pnas.1418031112

Bayne T, Seth AK, Massimini M, Shepherd J, Cleeremans A, Fleming SM, et al. Tests for consciousness in humans and beyond. Trends Cogn Sci. 2024;28(5):454–66. https://doi.org/10.1016/j.tics.2024.01.010

Bekoff M. The emotional lives of animals. Novato, CA: New World Library; 2007.

Ben-Haim MS, Monte OD, Fagan NA, Dunham Y, Hassin RR, Chang SWC, et al. Disentangling perceptual awareness from nonconscious processing in rhesus monkeys. P Natl Acad Sci USA. 2021;118(15). Article No. e2017543118. https://doi.org/10.1073/pnas.2017543118

Bentham J. Introduction to the principles of morals and legislation. 1828. Reprinted 1988, Prometheus, Amherst NY.

Bermond B. A neuropsychological and evolutionary approach to animal consciousness and animal suffering. Anim Welfare. 2001;10:S47–S62.

Birch J. Animal sentience and the precautionary principle. Animal Sentience. 2017;2(16); https://doi.org/10.51291/2377-7478.1200

Birch J. The edge of sentience: risk and precautions in humans, other animals, and AI. Oxford: Oxford University Press; 2024.

Birch J, Ginsburg S, Jablonka E. Unlimited Associative Learning and the origins of consciousness: a primer and some predictions. Biol Philos. 2020;35(6). Article No. 56. https://doi.org/10.1007/s10539-020-09772-0

Birch J, Schnell AK, Clayton NS. Dimensions of animal consciousness. Trends Cogn Sci. 2020;24(10):789–801. https://doi.org/10.1016/j.tics.2020.07.007

Bishop D. Rein in the four horsemen of irreproducibility. Nature. 2019;568 (7753):435.

Blackmore S, Troscianko ET. Consciousness. An introduction. 3rd ed. Abingdon: Routledge; 2018.

Block N. On a confusion about a function of consciousness. Behav Brain Sci. 1995;18(2):227–47. https://doi.org/10.1017/S0140525x00038188

Boly M, Seth AK, Wilke M, Ingmundson P, Baars B, Laureys S, et al. Consciousness in humans and non-human animals: recent advances and future directions. Front Psychol. 2013;4. Article No. 625. https://doi.org/10.3389/fpsyg.2013.00625

Braithwaite V. Do fish feel pain? Oxford: Oxford University Press; 2010.

Brette R. Brains as computers: metaphor, analogy, theory or fact? Front Ecol Evol. 2022;10. Article No. 878729. https://doi.org/10.3389/fevo.2022.878729

Bronfman ZZ, Ginsburg S, Jablonka E. The transition to minimal consciousness through the evolution of associative learning. Front Psychol. 2016;7. Article No. 1954. https://doi.org/10.3389/fpsyg.2016.01954

Broom DM. Sentience and animal welfare. Wallingford: CABI; 2014.

Budolphson M, Fischer B, Scovronick N. Animal welfare: methods to improve policy and practice. Science. 2023;381(6653):32–34. https://doi.org/10.1126/science.adi0121

Burghardt GM. Animal awareness - current perceptions and historical-perspective. Am Psychol. 1985;40(8):905–19. https://doi.org/10.1037/0003-066x.40.8.905

Butlin P. Sentience criteria to persuade the reasonable sceptic. Animal Sentience. 2022;7. https://doi.org/1051291/2377-7478.1741

Cabanac M, Cabanac AJ, Parent A. The emergence of consciousness in phylogeny. Behav Brain Res. 2009;198(2):267–72. https://doi.org/10.1016/j.bbr.2008.11.028

Cambridge Declaration on Consciousness. Francis Crick Memorial Conference, Cambridge; 2012. https://fcmconference.org/img/CambridgeDeclarationOnConsciousness.pdf

Cammaerts M-C, Cammaerts R. Are ants (Hymenoptera, Formicidae) capable of self recognition? J Sci. 2015;5:521–32. https://difusion.ulb.ac.be/vufind/Record/ULB-DIPOT:oai:dipot.ulb.ac.be.2013/219269/Details

Carruthers P. Human and animal minds. Oxford: Oxford University Press; 2019.

Chalmers DJ. Facing up to the problem of consciousness. J Consciousness Stud. 1995;2:200–19.

Chalmers DJ. Panpsychism and panprotopsychism. In: Alter, T. & Nagasawa, Y. Consciousness in the Physical World: Perspectives on Russellian Monism. New York: Oxford University Press; 2015, pp. 246–76.

Chamovitz D. What a plant knows. A field guide to the senses of your garden and beyond. New York: Oxford University Press; 2012.

Chang YP, Wang X, Wang JD, Wu Y, Yang LY, Zhu KJ, et al. A survey on evaluation of large language models. Acm T Intel Syst Tec. 2024;15(3). Article No. 39. https://doi.org/10.1145/3641289

Cheke LG, Clayton NS. Eurasian jays (*Garrulus glandarius*) overcome their current desires to anticipate two distinct future needs and plan for them appropriately. Biol Letters. 2012;8(2):171–75. https://doi.org/10.1098/rsbl.2011.0909

Chittka L. The mind of a bee. Princeton: Princeton University Press; 2022.

Clark RE, Squire LR. Classical conditioning and brain systems: the role of awareness. Science. 1998;280(5360):77–81. https://doi.org/10.1126/science.280.5360.77

Cleeremans A, Achoui D, Beauny A, Keuninckx L, Martin JR, Muñoz-Moldes S, et al. Learning to be conscious. Trends Cogn Sci. 2020;24(2):112–23. https://doi.org/10.1016/j.tics.2019.11.011

Cofré R, Herzog R, Mediano PAM, Piccinini J, Rosas FE, Perl YS, et al. Whole-brain models to explore altered states of consciousness from the bottom up. Brain Sci. 2020;10(9). Article No. 626. https://doi.org/10.3390/brainsci10090626

Colman AM, Pulford BD, Omtzigt D, al-Nowaihi A. Learning to cooperate without awareness in multiplayer minimal social situations. Cognitive Psychol. 2010;61(3):201–27. https://doi.org/10.1016/j.cogpsych.2010.05.003

Colpaert FC, Tarayre JP, Alliaga M, Slot LAB, Attal N, Koek W. Opiate self-administration as a measure of chronic nociceptive pain in arthritic rats. Pain. 2001;91(1–2):33–45. https://doi.org/10.1016/S0304-3959(00)00413-9

Crook RJ. Behavioral and neurophysiological evidence suggests affective pain experience in octopus. Iscience. 2021;24(3). Article No. 102229. https://doi.org/10.1016/j.isci.2021.102229

Crump A, Gibbons M, Barrett M, Birch J, Chittka L. Is it time for insect researchers to consider their subjects' welfare? Plos Biol. 2023;21(6). Article No. e3002138. https://doi.org/10.1371/journal.pbio.3002138

Damasio A. Descartes' error: emotion, reason and the human brain. New York: Putnams; 1994.

Darwin C. The expression of the emotions in man and animals. London: University of Chicago Press; 1872.

Davidson D. Mental events. In: Davidson D, editor. Essays on actions and events. Oxford: Oxford University Press; 1970. pp. 207–23.

Dawkins MS. From an animals point of view - motivation, fitness, and animal-welfare. Behav Brain Sci. 1990;13(1):1–60. https://doi.org/10.1017/S0140525x00077104

Dawkins MS. Through our eyes only? The search for animal consciousness. New York and Heidelberg: W.H.Freeman; 1993.

Dawkins MS. The science of animal sentience and the politics of animal welfare should be kept separate. Animal Sentience. 2022;31(8). https://doi.org/10.51291/2377-7478.170

Dehaene S. Consciousness and the brain: deciphering how the brain codes our thoughts. New York: Penguin Books; 2014.

Dehaene S, Changeux JP. Experimental and theoretical approaches to conscious processing. Neuron. 2011;70(2):200–27. https://doi.org/10.1016/j.neuron.2011.03.018

Dehaene S, Naccache L. Towards a cognitive neuroscience of consciousness: basic evidence and a workspace framework. Cognition. 2001;79(1-2):1–37. https://doi.org/10.1016/S0010-0277(00)00123-2

Dennett D. Consciousness explained. London: Little, Brown and Co.; 1991.

Descartes R. Meditations on first philosophy. Cambridge: Cambridge University Press; original publication 1637, reprinted 1996.

Descartes R. Passions of the soul. Indianapolis: Hackett; 1649.

Dettmer P. Immune: a journey into the mysterious system that keeps you alive. London: Hodder and Stoughton; 2021.

Devlin H. Introduction to fMRI 2012 [Available from: https://www.ndcn.ox.ac.uk/divisions/fmrib/what-is-fmri/introduction-t0-fmri]

Dickinson A. Associative learning and animal cognition. Philos T R Soc B. 2012;367(1603):2733–42. https://doi.org/10.1098/rstb.2012.0220

Diggles BK, Arlinghaus R, Browman HI, Cooke SJ, Cooper, Cowx IG. Reasons to be skeptical about sentience and pain in fishes and aquatic invertebrates. Rev Fish Sci Aquaculture. 2024;32:127–50. https://doi.org/10.1080/23308249.2023.2257802

Drissi I, Woods WA, Woods CG. Understanding the genetic basis of congenital insensitivity to pain. Br Med Bull. 2020;133(1):65–78. https://doi.org/10.1093/bmb/ldaa003

Droege P, Weiss DJ, Schwob N, Braithwaite V. Trace conditioning as a test for animal consciousness: a new approach. Anim Cogn. 2021;24(6):1299–304. https://doi.org/10.1007/s10071-021-01522-3

Dung L. Assessing tests of animal consciousness. Conscious Cogn. 2022;105: 103410. https://doi.org/10.1016/j.concog.2022.103410

Dylla KV, Galili DS, Szyszka P, Lüdke A. Trace conditioning in insects-keep the trace! Front Physiol. 2013;4. Article No. 67. https://doi.org/10.3389/fphys.2013.00067

Eagles DA, Chow CY, King GF. Fifteen years of Na 1.7 channels as an analgesic target: why has excellent in vitro pharmacology not translated into in vivo analgesic efficacy? Brit J Pharmacol. 2022;179(14):3592–611. https://doi.org/10.1111/bph.15327

Eccles JC. Evolution of consciousness. P Natl Acad Sci USA. 1992;89(16):7320–24. https://doi.org/10.1073/pnas.89.16.7320

Elwood RW. Caution is required when considering sentience in animals: a response to the commentary by Briffa (2022) on "Hermit crabs, shells, and sentience". Anim Cogn. 2022;25(6):1371–74. https://doi.org/10.1007/s10071-022-01655-z

Elwood RW, Appel M. Pain experience in hermit crabs? Anim Behav. 2009; 77(5):1243–46. https://doi.org/10.1016/j.anbehav.2009.01.028

Emery NJ, Clayton NS. Effects of experience and social context on prospective caching strategies by scrub jays. Nature. 2001;414(6862):443–46. https://doi.org/10.1038/35106560

Farisco M, Changeux JP. About the compatibility between the perturbational complexity index and the global neuronal workspace theory of consciousness. Neurosci Conscious. 2023;2023(1). Article No. niad016. https://doi.org/10.1093/nc/niad016

Freeman WJ, Watts JW. Psychosurgery in the treatment of mental disorders and intractable pain. Springfield, IL: Thomas; 1950.

Frith C, Rees G. A brief history of the scientific approach to the study of consciousness. In: M. Velmans and S Schneider, editors. The Blackwell companion to consciousness. Oxford: Blackwell Publishing; 2007. pp. 72–78.

Fuss T, Witte K. Sex differences in color discrimination and serial reversal learning in mollies and guppies. Curr Zool. 2019;65(3):323–32. https://doi.org/10.1093/cz/zoz029

Gallup GG. Chimpanzees. Self-recognition. Science. 1970;167(3914):86. https://doi.org/10.1126/science.167.3914.86

Gallup GG, Anderson JR. Self-recognition in animals: where do we stand 50 years later? Lessons from cleaner wrasse and other species. Psychol Conscious. 2020;7(1):46–58. https://doi.org/10.1037/cns0000206

Ganin Y, Ustinova E, Ajakan H, Germain P, Larochelle H, Laviolette F, et al. Domain-adversarial training of neural networks. J Mach Learn Res. 2016;17. Article No. 59.

Gerber B, Yarali A, Diegelmann S, Wotjak CT, Pauli P, Fendt M. Pain-relief learning in flies, rats, and man: basic research and applied perspectives. Learn Memory. 2014;21(4):232–52. https://doi.org/10.1101/lm.032995.113

Ghosh P. Are animals conscious?: BBC; 2024 [Available from: https://www.bbc.co.uk/search?q=animals+conscious&seqId=7125ffc0-2d3f-11ef-ad23-5fc4b7343564&d=HOMEPAGE_PS].

Gibbons M, Pasquini E, Kowalewska A, Read E, Gibson S, Crump A, et al. Noxious stimulation induces self-protective behavior in bumblebees. Iscience. 2024;27(8). Article No. 110440. https://doi.org/10.1016/j.isci.2024.110440

Gibbons M, Sarlak S, Chittka L. Descending control of nociception in insects? P Roy Soc B-Biol Sci. 2022;289(1978). Article No. 20220599. https://doi.org/10.1098/rspb.2022.0599

Ginsburg S, Jablonka E. The evolution of the sensitive soul: learning and the origins of consciousness. Cambridge, MA: MIT Press; 2019.

Glanzman DL. The cellular basis of classical conditioning in *Aplysia californica* – it's less simple than you think. Trends Neurosci. 1995;18(1):30–36. https://doi.org/10.1016/0166-2236(95)93947-V

Godfrey-Smith P. Other minds: the octopus and the evolution of intelligent life. London: William Collins; 2016.

Goldacre B. Bad science. London: Harper Collins; 2008.

Goldberg YP, MacFarlane J, MacDonald ML, Thompson J, Dube MP, Mattice M, et al. Loss-of-function mutations in the Na 1.7 gene underlie congenital indifference to pain in multiple human populations. Clin Genet. 2007;71(4):311–19. https://doi.org/10.1111/j.1399-0004.2007.00790.x

Goodall J. Forward. The emotional lives of animals. Novato, CA: New World Library; 2007. pp. xi–xv.

Grabenhorst F, Rolls ET, Parris BA, d'Souza AA. How the brain represents the reward value of fat in the mouth. Cereb Cortex. 2010;20(5):1082–91. https://doi.org/10.1093/cercor/bhp169

Granger CWJ. Investigating causal relations by econometric models and cross-spectral methods. Econometrica. 1969;37(3):424–38. https://doi.org/10.2307/1912791

Green DM, Birdsall TG, Tanner WP. Signal detection as a function of signal intensity and duration. J Acoust Soc Am. 1957;29(4):523–31. https://doi.org/10.1121/1.1908951

Griffin D. The question of animal awareness. New York: Rockefeller University Press; 1976.

Griffin D. Animal thinking. Harvard: Harvard University Press; 1984.

Gutierrez T, Crystal JD, Zvonok AM, Makriyannis A, Hohmann AG. Self-medication of a cannabinoid CB agonist in an animal model of neuropathic pain. Pain. 2011;152(9):1976–87. https://doi.org/10.1016/j.pain.2011.03.038

Hampton RR. Rhesus monkeys know when they remember. P Natl Acad Sci USA. 2001;98(9):5359–62. https://doi.org/10.1073/pnas.071600998

Harford T. How to make the world add up. 2020. London: Little, Brown Book Group Ltd.

Harnad S. Insentient "cognition"? Animal Sentience. 2023;33. https://doi.org/10.51291/2377-7478.1780

Harrison R. Animal machines: the new factory farming industry. London: Vincent Stuart; 1964.

Hart PJB. Exploring the limits to our understanding of whether fish feel pain. J Fish Biol. 2023;102(6):1272–80. https://doi.org/10.1111/jfb.15386

Havlík M, Kozáková E and Horácek J. Why and how: the future of the central questions of consciousness. Front. Psychol. 2017; 8:1797. https://doi.org/10.3389/fpsyg.2017.01797

Hughes BO, Black AJ. Preference of domestic hens for different types of battery cage floor. Brit Poultry Sci. 1973;14(6):615–19. https://doi.org/10.1080/00071667308416071

Huis A von. Welfare of farmed insects. J Insects Food Feed. 2021;7:573–84. https://doi.org/10.3920/JIFF2020.0061

Humphrey N. Sentience: the invention of consciousness. Oxford: Oxford University Press; 2022.

Huxley TH. On the hypothesis that animals are automata, and its history. Collected essays: method and results. 1. London: Macmillan; 1874 (Reprinted 1893). pp. 195–250.

Jékely G, Godfrey-Smith P, Keijzer F. Reafference and the origin of the self in early nervous system evolution. Philos T R Soc B. 2021;376(1821). Article No. 20190764. https://doi.org/10.1098/rstb.2019.0764

Jones G, Teeling EC. The evolution of echolocation in bats. Trends Ecol Evol. 2006;21(3):149–56. https://doi.org/10.1016/j.tree.2006.01.001

Kahneman D. Thinking fast and slow. Penguin; 2011.

Kakrada E, Colombo M. Mirror mirror on the wall, it's not the mark I care about at all. Learn Motiv. 2022;77. Article No. 101785. https://doi.org/10.1016/j.lmot.2022.101785

Kasneci E, Sessler K, Küchemann S, Bannert M, Dementieva D, Fischer F, et al. ChatGPT for good? On opportunities and challenges of large language models for education. Learn Individ Differ. 2023;103. Article No. 102274. https://doi.org/10.1016/j.lindif.2023.102274

Kastrup B. The universe in consciousness. J Consciousness Stud. 2018;25(5-6):125–55.

Key B. Fish do not feel pain and its implications for understanding phenomenal consciousness. Biol Philos. 2015;30(2):149–65. https://doi.org/10.1007/s10539-014-9469-4

Kim J. Philosophy of mind. 3rd ed. Colorado: Westview Press; 2011.

King JR, Gramfort A, Schurger A, Naccache L, Dehaene S. Two distinct dynamic modes subtend the detection of unexpected sounds. Plos One. 2014;9(1). Article No. e85791. https://doi.org/10.1371/journal.pone.0085791

Koch C. The quest for consciousness. A neurobiological approach. Englewood CO: Roberts and Co.; 2004.

Koch C, Massimini M, Boly M, Tononi G. Neural correlates of consciousness: progress and problems. Nat Rev Neurosci. 2016;17(6):395. https://doi.org/10.1038/nrn.2016.61

Kohda M, Hotta T, Takeyama T, Awata S, Tanaka H, Asai JY, et al. If a fish can pass the mark test, what are the implications for consciousness and self-awareness testing in animals? Plos Biol. 2019;17(2). Article No. e3000021. https://doi.org/10.1371/journal.pbio.3000021

Kuroda T, Mizutani Y, Cançado CRX, Podlesnik CA. Reversal learning and resurgence of operant behavior in zebrafish. Behav Process. 2017;142:79–83. https://doi.org/10.1016/j.beproc.2017.06.004

Lamme VAF. Why visual attention and awareness are different. Trends Cogn Sci. 2003;7(1):12–18. https://doi.org/10.1016/S1364-6613(02)00013-X

LeDoux JE. Coming to terms with fear. P Natl Acad Sci USA. 2014;111(8):2871–78. https://doi.org/10.1073/pnas.1400335111

LeDoux JE, Hofmann SG. The subjective experience of emotion: a fearful view. Curr Opin Behav Sci. 2018;19:67–72. https://doi.org/10.1016/j.cobeha.2017.09.011

LeDoux JE, Pine DS. Using neuroscience to help understand fear and anxiety: a two-system framework. Am J Psychiat. 2016;173(11):1083–93. https://doi.org/10.1176/appi.ajp.2016.16030353

Levin, J. Functionalism. Stanford Encyclopedia of Philosophy. 2004. https://plato.stanford.edu/entries/functionalism

Liu YX, Day LB, Summers K, Burmeister SS. Learning to learn: advanced behavioural flexibility in a poison frog. Anim Behav. 2016;111:167–72. https://doi.org/10.1016/j.anbehav.2015.10.018

Logothetis NK, Pauls J, Augath M, Trinath T, Oeltermann A. Neurophysiological investigation of the basis of the fMRI signal. Nature. 2001;412(6843):150–57. https://doi.org/10.1038/35084005

Luppi AI, Uhrig L, Tasserie J, Signorelli CM, Stamatakis EA, Destexhe A, et al. Local orchestration of distributed functional patterns supporting loss and restoration of consciousness in the primate brain. Nat Commun. 2024;15(1). Article No. 2171. https://doi.org/10.1038/s41467-024-46382-w

Lynn B. The fibre composition of cutaneous nerves and the classification and response properties of cutaneous afferents, with particular reference to nocioception. Pain Rev. 1994;1:172–83.

Macphail EM. The comparative psychology of intelligence. Behav Brain Sci. 1987;10(4):645–56. https://doi.org/10.1017/S0140525x00054984

Mallatt J, Feinberg TE. Multiple routes to animal consciousness: constrained multiple realizability rather than modest identity theory. Front Psychol. 2021;12. Article No. 732336. https://doi.org/10.3389/fpsyg.2021.732336

Margulis L. The conscious cell. Cajal and Consciousness. Ann Ny Acad Sci. 2001;929:55–70. https://doi.org/10.1111/j.1749-6632.2001.tb05707.x

Martinez J, von Nolting, C. Review: "Animal welfare"- a European concept. Animal. 2023;17. Article No. 100839. https://doi.org/10.1016/j.animal.2023.100839

Mashour GA. Anesthesia and the neurobiology of consciousness. Neuron. 2024;112(10):1553–67. https://doi.org/10.1016/j.neuron.2024.03.002

Mashour GA, Alkire MT. Evolution of consciousness: Phylogeny, ontogeny, and emergence from general anesthesia. P Natl Acad Sci USA. 2013;110:10357–64. https://doi.org/10.1073/pnas.1301188110

Mason GJ, Lavery JM. What is it like to be a bass? Red herrings, fish pain and the study of animal sentience. Front Vet Sci. 2022;9. Article No. 948567. https://doi.org/10.3389/fvets.2022.948567

McCormick PA. Orienting attention without awareness. J Exp Psychol Human. 1997;23(1):168–80. https://doi.org/10.1037/0096-1523.23.1.168

Mellor DJ. Welfare-aligned sentience: enhanced capacities to experience, interact, anticipate, choose and survive. Animals-Basel. 2019;9(7). Article No. 440. https://doi.org/10.3390/ani9070440

Merker B. The liabilities of mobility: a selection pressure for the transition to consciousness in animal evolution. Conscious Cogn. 2005;14(1):89–114. https://doi.org/10.1016/S1053-8100(03)00002-3

Midgley M. Animals and why they matter. A journey round the species barrier. Harmondsworth: Pelican Books; 1983.

Mogil JS, Yu L, Basbaum AI. Pain genes?: Natural variation and transgenic mutants. Annu Rev Neurosci. 2000;23:777–811. https://doi.org/10.1146/annurev.neuro.23.1.777

Montero B. Does bodily awareness interfere with highly skilled movement? Inquiry. 2010;53(2):105–22. https://doi.org/10.1080/00201741003612138

Morrell V. Animal wise: the thoughts and emotions of our fellow creatures. New York: Crown Publishing Group; 2013.

Mudrik L, Boly M, Dehaene S, Fleming SM, Lamme V, Seth A, Melloni L. Unpacking the complexities of consciousness: theories and reflections. Neurosci Biobehav Rev. 2025; 170. Article No. 106053. https://doi.org/10.1016/j.neubiorev.2025.106053

Nagel T. What is it like to be a bat. Philos Rev. 1974;83(4):435–50. https://doi.org/10.2307/2183914

Navratilova E, Xie JY, King T, Porreca F. Evaluation of reward from pain relief. Addiction Rev. 2013;1282:1–11. https://doi.org/10.1111/nyas.12095

New York Declaration. The New York Declaration on Animal Consciousness 2024. https://sites.google.com/nyu.edu/nydeclaration/declaration.

Newell BR, Shanks DR. Unconscious influences on decision making: a critical review. Behav Brain Sci. 2014;37(1):88–108. https://doi.org/10.1017/S0140525x12003214

Nilsen AS, Juel BE, Thürer B, Aamodt A, Storm JF. Are we really unconscious in "unconscious" states? Common assumptions revisited. Front Hum Neurosci. 2022;16. Article No. 987051. https://doi.org/10.3389/fnhum.2022.987051

O'Leary MA, Bloch JI, Flynn JJ, Gaudin TJ, Giallombardo A, Giannini NP, et al. The placental mammal ancestor and the post-K-Pg radiation of placentals. Science. 2013;339(6120):662–67. https://doi.org/10.1126/science.1229237

Ossipov MH, Morimura K, Porreca F. Descending pain modulation and chronification of pain. Curr Opin Support Pa. 2014;8(2):143–51. https://doi.org/10.1097/SPC.0000000000000055

Panksepp J. Affective neuroscience: the foundation of human and animal emotion. Oxford: Oxford University Press; 1998.

Parker MO, Gaviria J, Haigh A, Millington ME, Brown VJ, Combe FJ, et al. Discrimination reversal and attentional sets in zebrafish. Behav Brain Res. 2012;232(1):264–68. https://doi.org/10.1016/j.bbr.2012.04.035

Passingham RE. Understanding the pre-frontal cortex: selective advantage, connectivity and neural operations. Oxford: Oxford University Press; 2021.

Pennartz CMA, Farisco M, Evers K. Indicators and criteria of consciousness in animals and intelligent machines: an inside-out approach. Front Syst Neurosci. 2019;13. Article No. 25. https://doi.org/10.3389/fnsys.2019.00025

Perrett DI, Rolls ET, Caan W. Visual neurones responsive to faces in the monkey temporal cortex. Exp Brain Res. 1982;47(3):329–42. https://doi.org/10.1007/BF00239352

Pham TM, Hagman B, Codita A, Van Loo PL, Strommer L, Baumans V. Housing environment influences the need for pain relief during post-operative recovery in mice. Physiol Behav. 2010;99(5):663–68. https://doi.org/10.1017/S0140525X00076512

Polger TW. Natural minds. MIT Press; 2006.

Porreca F, Navratilova E. Reward, motivation, and emotion of pain and its relief. Pain. 2017;158(4):S43–S9. https://doi.org/10.1097/j.pain.0000000000000798

Premack D, Woodruff G. Does the chimpanzee have a theory of mind. Behav Brain Sci. 1978;1(4):515–26. https://doi.org/10.1097/j.pain.0000000000000798

Raichie M. Functional brain imaging and human brain function. J Neurosci. 2003;23:3959–62. https://doi.org/10.1523/JNEUROSCI.23-10-03959.2003

Rainville P, Carrier B, Hofbauer RK, Bushnell MC, Duncan GH. Dissociation of sensory and affective dimensions of pain using hypnotic modulation. Pain. 1999;82(2):159–71. https://doi.org/10.1016/S0304-3959(99)00048-2

Rayburn-Reeves RM, Laude JR, Zentall TR. Pigeons show near-optimal win-stay/lose-shift performance on a simultaneous-discrimination, midsession reversal task with short intertrial intervals. Behav Process. 2013;92:65–70. https://doi.org/10.1016/j.beproc.2012.10.011

Reed N, McLeod P, Dienes Z. Implicit knowledge and motor skill: what people who know how to catch don't know. Conscious Cogn. 2010;19(1):63–76. https://doi.org/10.1016/j.concog.2009.07.006

Roige A. There is an epistemic problem in animal consciousness research. Phenomenol Cogn Sci. 2023;1. Article No. 3. https://doi.org/10.1007/s11097-023-09912-3

Rollin B. The unheeded cry. Oxford: Oxford University Press; 1989.

Rolls ET. Consciousness absent and present: a neurophysiological exploration. Prog Brain Res. 2004;144:95–106. https://doi.org/10.1016/S0079-6123(03)14406-8

Rolls ET. Invariant visual object and face recognition: neural and computational bases, and a model, VisNet. Front Comput Neurosc. 2012;6. Article No. 35. https://doi.org/10.3389/fncom.2012.00035

Rolls ET. Emotion and decision-making explained. Oxford: Oxford University; 2014.

Rolls ET. Neural computations underlying phenomenal consciousness: a higher order syntactic thought theory. Front Psychol. 2020;11:655. https://doi.org/10.3389/fpsyg.2020.00655

Rolls ET. A neuroscience level of explanation approach to the mind and the brain. Front Comput Neurosc. 2021;15. Article No. 649679. https://doi.org/10.3389/fncom.2021.649679

Rolls ET. Emotion, motivation, decision-making, the orbitofrontal cortex, anterior cingulate cortex, and the amygdala. Brain Struct Funct. 2023;228(5):1201–57. https://doi.org/10.1007/s00429-023-02644-9

Rolls ET, O'Doherty J, Kringelbach ML, Francis S, Bowtell R, McGlone F. Representations of pleasant and painful touch in the human orbitofrontal and cingulate cortices. Cereb Cortex. 2003;13(3):308–17. https://doi.org/10.1093/cercor/13.3.308

Rose JD, Arlinghaus R, Cooke SJ, Diggles BK, Sawynok W, Stevens ED, et al. Can fish really feel pain? Fish Fish. 2014;15(1):97–133. https://doi.org/10.1111/faf.12010

Rosenthal D. Thinking that one thinks. In: Humphreys MDaGW, editor. Consciousness. Oxford: Blackwell Publishing; 1993. pp. 197–223.

Rosenthal D. Consciousness and mind. Oxford: Oxford University Press; 2005.

Rowan A, N., D'Silva, J. Duncan, I.J.H., Palmer, N. Animal sentience - history, science and politics. Animal Sentience. 2021;31. https://doi.org/10.51291/2377-7478.1697

Ryle G. The concept of mind. London: Hutchinson; 1949.

Sanders RD, Tononi G, Laureys S, Sleigh JW. Unresponsiveness ≠ Unconsciousness. Anesthesiology. 2012;116(4):946–59. https://doi.org/10.1097/ALN.0b013e318249d0a7

Schelonka K, Graulty C, Canseco-Gonzalez E, Pitts MA. ERP signatures of conscious and unconscious word and letter perception in an inattentional blindness paradigm. Conscious Cogn. 2017;54:56–71. https://doi.org/10.1016/j.concog.2017.04.009

Schräder J, Habel U, Jo HG, Walter F, Wagels L. Identifying the duration of emotional stimulus presentation for conscious versus subconscious perception via hierarchical drift diffusion models. Conscious Cogn. 2023;110:103493. https://doi.org/10.1016/j.concog.2023.103493

Segundo-Ortin M, Calvo, P. Plant sentience? Between romanticism and denial: science. Animal Sentience. 2023;33(1). https://doi.org/10.51291/2377-7478.1772

Seth AK, Baars BJ, Edelman DB. Criteria for consciousness in humans and other mammals. Conscious Cogn. 2005;14(1):119–39. https://doi.org/10.1016/j.concog.2004.08.006

Seth AK, Bayne T. Theories of consciousness. Nat Rev Neurosci. 2022;23(7):439–52. https://doi.org/10.1038/s41583-022-00587-4

Shapiro L. Elisabeth, Princess of Bohemia. The Stanford Encyclopedia of Philosophy 2013. https://plato.stanford.edu/entries/elisabeth-bohemia/

Sherrington C.S. Integrative action of the nervous system. New Haven CT: Yale University Press; 1900.

Shettleworth SJ. Clever animals and killjoy explanations in comparative psychology. Trends Cogn Sci. 2010;14(11):477–81. https://doi.org/10.1016/j.tics.2010.07.002

Shevlin H. How could we know when a robot was a moral patient? Camb Q Healthc Ethic. 2021;30(3):459–71. https://doi.org/10.1017/S0963180120001012

Singer P. Animal liberation. London: Jonathan Cape; 1975.

Sitt JD, King JR, El Karoui I, Rohaut B, Faugeras F, Gramfort A, et al. Large scale screening of neural signatures of consciousness in patients in a vegetative or minimally conscious state. Brain. 2014;137:2258–70. https://doi.org/10.1093/brain/awu141

Sklar AY, Levy N, Goldstein A, Mandel R, Maril A, Hassin RR. Reading and doing arithmetic nonconsciously. P Natl Acad Sci USA. 2012;109(48):19614–19. https://doi.org/10.1073/pnas.1211645109

Smart JJC. The mind/brain identity theory. In: Zalth EN, editor. The Stanford Encylcopedia of Philosophy. 2000. https://plato.stanford.edu/entries/mind-identity/

Sneddon LU. Pain in aquatic animals. J Exp Biol. 2015;218(7):967–76. https://doi.org/10.1242/jeb.088823

Sneddon LU. Comparative physiology of nociception and pain. Physiology. 2018;33(1):63–73. https://doi.org/10.1152/physiol.00022.2017

Sneddon LU. Evolution of nociception and pain: evidence from fish models. Philos T R Soc B. 2019;374(1785). Article No. 20190290. https://doi.org/10.1098/rstb.2019.0290

Stoljar D. Physicalism: Stanford University; 2016 https://plato.stanford.edu/entries/physicalism/#pagetopright

Suddendorf T, Butler DL. The nature of visual self-recognition. Trends Cogn Sci. 2013;17(3):121–27. https://doi.org/10.1016/j.tics.2013.01.004

Swets JA, Tanner WP, Birdsall TG. Decision-processes in perception. Psychol Rev. 1961;68(5):301–40. https://doi.org/10.1037/h0040547

Taschereau-Dumouchel V, Michel M, Lau H, Hofmann SG, LeDoux JE. Putting the "mental" back in "mental disorders": a perspective from research on fear and anxiety. Mol Psychiatr. 2022;27(3):1322–30. https://doi.org/10.1038/s41380-021-01395-5

Technical University of Munich. How a robot can recognize itself. 2019. https://www.youtube.com/watch?v=s7xy3aHaytQ)

Thom JM, Clayton NS. Re-caching by western scrub-jays cannot be attributed to stress. Plos One. 2013;8(1). Article No. e52936. https://doi.org/10.1371/journal.pone.0052936

Thorpe WH. The assessment of pain and d istress in animals. London: Her Majesty's Stationery Office (HMSO); 1965. pp. 71–79.

Tinbergen N. The study of instinct. Oxford: Oxford University Press; 1951.

Travers E, Frith CD, Shea N. Learning rapidly about the relevance of visual cues requires conscious awareness. Q J Exp Psychol. 2018;71(8):1698–713. https://doi.org/10.1080/17470218.2017.1373834

Tye M. Tense bees and shell-shocked crabs: are animals conscious. Oxford: Oxford University Press; 2017.

van der Vaart E, Verbrugge R, Hemelrijk CK. Corvid re-caching without 'theory of mind': a model. Plos One. 2012;7(3). Article No. e32904. https://doi.org/10.1371/journal.pone.0032904

Vimal, RLP. Subjective experience aspects of consciousness. Part 1. Integration of classical, quantum, and subquantum concepts. Neuroquantology 2009;7(3): 390–410. ISSN 1303 5150.

Watson JB. Psychology as the behaviorist views it. Psychol Rev. 1913;20:158–77.

Watson JB. Psychology from the standpoint of a behaviorist. Philadelphia: Lippincott; 1929.

Weiskrantz L. Consciousnes lost and found. Oxford: Oxford University Press; 1997.

Weizenbaum J. Eliza - a computer program for study of natural language communication between man and machine. Commun ACM. 1966;9(1):36–46. https://doi.org/10.1145/365153.365168

Wenstrup JJ, Portfors CV. Neural processing of target distance by echolocating bats: Functional roles of the auditory midbrain. Neurosci Biobehav R. 2011;35(10):2073–83. https://doi.org/10.1016/j.neubiorev.2010.12.015

Willingham DB, Salidis J, Gabrieli JDE. Direct comparison of neural systems mediating conscious and unconscious skill learning. J Neurophysiol. 2002;88(3):1451–60. https://doi.org/10.1152/jn.2002.88.3.1451

Woodward J. Making things happen: a theory of causal explanation. Oxford: Oxford University Press; 2005.

Yong E. An immense world. How animal senses reveal the hidden realms sound us. London: Penguin Random House; 2023.

Zentall TR. Selective and divided attention in animals. Behav Process. 2005;69(1):1–15. https://doi.org/10.1016/j.beproc.2005.01.004

Index

For the benefit of digital users, indexed terms that span two pages (e.g., 52–53) may, on occasion, appear on only one of those pages.